Lecture Notes in Mathematics

Edited by A. Dold and B. Eckmann

444

T0225959

F. van Oystaeyen

Prime Spectra in Non-Commutative Algebra

Springer-Verlag
Berlin · Heidelberg · New York 1975

Prof. Freddy M. J. van Oystaeyen
Departement Wiskunde
Universiteit Antwerpen
Universiteitsplein 1
2610 Wilrijk/Belgium

Library of Congress Cataloging in Publication Data

Oystaeyen, F van, 1947-
Prime spectra in noncommutative algebra.

(Lecture notes in mathematics ; 444)
Bibliography: p.
Includes index.
1. Associative algebras. 2. Associative rings.
3. Modules (algebra) 4. Ideals (algebra) 5. Sheaves, theory of. I. Title. II. Series: Lecture notes in mathematics (Berlin) ; 444.
QA3.L28 no. 444 [QA251.5] 510'.8s [512'.24]
75-4877

AMS Subject Classifications (1970): 14A20, 16-02, 16A08, 16A12, 16A16, 16A40, 16A46, 16A64, 16A66, 18F20

ISBN 3-540-07146-6 Springer-Verlag Berlin · Heidelberg · New York
ISBN 0-387-07146-6 Springer-Verlag New York · Heidelberg · Berlin

This work is subject to copyright. All rights are reserved, whether the whole or part of the material is concerned, specifically those of translation, reprinting, re-use of illustrations, broadcasting, reproduction by photocopying machine or similar means, and storage in data banks.

Under § 54 of the German Copyright Law where copies are made for other than private use, a fee is payable to the publisher, the amount of the fee to be determined by agreement with the publisher.

© by Springer-Verlag Berlin · Heidelberg 1975. Printed in Germany.

Offsetdruck: Julius Beltz, Hemsbach/Bergstr.

CONTENTS

LEITFADEN

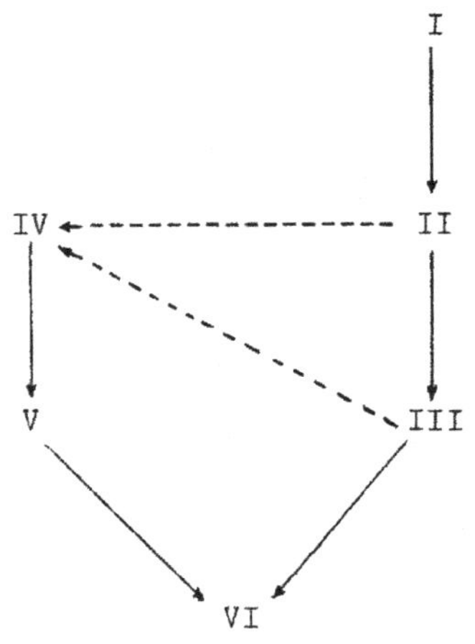

ACKNOWLEDGEMENT

The author is indebted to Professor Dave Murdoch at the University of British Columbia for reading the manuscript and for many helpful suggestions.

Part of this research was done at Cambridge University and I am thankful to the people of the mathematical department for the hospitality enjoyed there.

I was able to continue with this work while serving in the Belgian army and I thank my superiors for not trying too hard to make a soldier out of a mathematician.

I am obliged to the University of Antwerp for facilities and the necessary financial support.

I thank Melanie for reading the illegible and for careful typing.

SYMMETRIC LOCALIZATION AND SHEAVES

Introduction

Present notes are mainly concerned with two topics : symmetric localization and pseudo-places of algebras over fields. The first three sections deal with localization theory while in the remaing sections the accent is on pseudo-places. This split up is reflected in a shift of interest from prime ideals to completely prime ideals.

After a brief summary of the basic facts about kernel functors, localization techniques expounded by P. Gabriel and O. Goldman are adapted, in Section II, so as to yield a satisfactory ideal theory. In Section III we construct a presheaf of noncommutative rings on the prime spectrum, Spec R, of a left Noetherian ring. If R is a prime ring then this presheaf is a sheaf and Spec R is then said to be an affine scheme. Although Spec is not necessarily functorial, studying it is worthwile because of the many local properties that still hold. For example, if A is an ideal of R such that the Zariski open subset X_A of Spec R is such that the associated localization functor Q_A has property (T) discussed by O. Goldman, then X_A is an affine scheme too, in fact $X_A \cong \text{Spec } Q_A(R)$, where $Q_A(R)$ is the ring of quotients with respect to Q_A. Local properties like this are related to the question whether the extension of an ideal A of R to a left ideal $Q_\sigma(A)$ of $Q_\sigma(R)$, for some symmetric kernel functor σ, is also an ideal of $Q_\sigma(R)$. Therefore the local properties of the sheaf Spec R lean heavily on the ideal theory expounded in II. 1.

The relation between symmetric localization at a prime ideal of a left Noetherian ring and the J. Lambek - G. Michler torsion theory, cf. [21], is explained in II. 2. It turns out that the symmetric kernel functor σ_{R-P} at a prime ideal P of R is the biggest symmetric kernel functor smaller than the J. Lambek -G. Michler torsion theory σ_P. Inspired by this we define a quasi-prime kernel functor, in II. 2., to be the biggest symmetric functor smaller than the kernel functor induced by a quasi-supporting module. The latter concept is strongly related to Goldman's supporting modules. It seems to be natural to consider a quasi-prime as a "generalization" of a prime ideal while a prime kernel functor in the sense of Goldman [12] should then be the analogue of a prime left ideal. It should be noted that the correspondence between prime ideals and quasi-prime functors is the better one because to every prime ideal there does indeed correspond a quasi-prime and moreover, in some special cases, the correspondence may be described completely.

Using completely prime ideals we get a generalization of the sheaf $Gam_k A$ on a k-algebra A, introduced by I.G. Connell. Therefore we consider pseudo-places of algebras over fields, which are ring homomorphisms of certain subrings of the k-algebra A, thus generalizing the concept of a place of a field. The general theory of pseudo-places and their specialization applies so as to yield information about primes in algebras over fields, allowing us to construct a sheaf $Prim_k(A)$ on A. In the commutative case $Prim_k$ is a functor $\underline{Alg}_k \to \underline{Sheaves}$, but this is not always true in the noncommutative case. The kernel functors at the stalks of $Prim_k(A)$ turn out to be the quasi-primes associated to prime ideals. However there is added interest to localization at primes because there are several symmetric kernel functors which may be associated to a prime. Interrelations between these are the main subject of IV. 5..

A further application of the theory of pseudo-places is given in Section V. To every finite abelian group G we construct a functor $\mathcal{D}(G)$ from fields and surjective places to crossed product skew fields and galoisian pseudo-places. The skew field $\mathcal{D}_k(G)$ ($\mathcal{D}_k(G) = \mathcal{D}(G)(k)$) has a generic property :

every crossed product $(G,l/k,\{C_{\sigma,\tau}\})$, where l/k is an abelian extension with $Gal(l/k) \cong G$, defined by a symmetric factor set $\{C_{\sigma,\tau} | \sigma,\tau \in G\}$, is residue algebra of $\mathcal{D}_k(G)$ under a galoisian pseudo-place. In this way a parametrization of certain subgroups of the Brauer Group is obtained.

In the Appendix, Section VI, the theory of symmetric localization is applied to an Azumaya algebra R with center C. It will be proved there that symmetric kernel functors on M(R) may be "descended" to kernel functors on M(C) when they have property (T). Moreover, localization at a prime ideal P of R has all the good properties one could hope for and therefore Spec R is as close as one can get to an affine sheaf in the commutative case.
The J. Lambek, G. Michler torsion theory σ_P at P coincides with the symmetric σ_{R-P} in case R is a left Noetherian Azumaya algebra, and σ_P has property (T). The ring of quotients $Q_\sigma(R)$ at a symmetric kernel functor σ is an Azumaya algebra and it is a central extension of R, so if σ is a T-functor then every ideal $A \notin T(\sigma)$ has the property that $Q_\sigma(A)$ is an ideal of $Q_\sigma(R)$.
Localization of Azumaya algebras over valuation rings is closely related to the theory of unramified pseudo-places of central simple algebras.

I. GENERALITIES ON LOCALIZATION

I. 1. Kernel Functors.

All rings considered are assumed to have a unit element. Module will mean left module and all modules will be unitary. Ideal stands for two-sided ideal. Let R be a ring. Denote by M(R) the category of R-modules. A functor σ from M(R) to M(R) is a kernel functor if it has the following properties :

1. For every $M \in M(R)$, $\sigma(M)$ is a submodule of M.
2. If $f \in \mathrm{Hom}_R(M,N)$, then $f(\sigma(M)) \subset \sigma(N)$.
3. For any submodule N of M we have $\sigma(N) = N \cap \sigma(M)$.

Let σ be a kernel functor then $M \in M(R)$ is said to be σ-torsion if $\sigma(M) = M$, it is σ-torsion free if $\sigma(M) = 0$. A torsion theory is in fact defined by giving the class of torsion objects and the class of torsion free objects. A kernel functor σ on M(R) is called idempotent if $\sigma(M/\sigma(M)) = 0$ for all $M \in M(R)$. To an arbitrary kernel functor σ there is associated a filter $T(\sigma)$ consisting of left ideals A of R such that R/A is a σ-torsion module. The filter $T(\sigma)$, sometimes called a topology, has the following properties :

1. If $A \in T(\sigma)$ and if B is a left ideal of R such that $A \subset B$ then $B \in T(\sigma)$.
2. If $A,B \in T(\sigma)$ then $A \cap B \in T(\sigma)$.
3. For every $A \in T(\sigma)$ and any $x \in R$ there exists a $B \in T(\sigma)$ such that $B\,x \subset A$.
4. Let $M \in M(R)$, then $x \in \sigma(M)$ if and only if there is an $A \in T(\sigma)$ such that $A\,x = 0$.

Any filter T of left ideals of R, having properties 1,2,3, listed above, defines a topology in R such that R becomes a topological ring. Conversely, to such a filter T there corresponds a functor σ on M(R),

defined by $\sigma(M) = \{x \in M, A\,x = 0$ for some $A \in T\}$. This σ is a kernel functor with $T(\sigma) = T$. The one-to-one correspondence between kernel functors and topologies of the prescribed type, makes it possible to talk about the set $F(R)$ of kernel functors on $M(R)$. A $\sigma \in F(R)$ induces a topology in every $M \in M(R)$ taking for the neighborhoods of 0 in M the submodules N of M for which the quotient module M/N is σ-torsion. This topology is called the <u>σ-topology</u> in M.
Taking $M = R$, the σ-topology induced in R is exactly $T(\sigma)$. The set $F(R)$ may be partially ordered by putting $\sigma \leqslant \tau$ if and only if $\sigma(M) \subset \tau(M)$ for all $M \in M(R)$. Gathering results from [12] we obtain the following :

<u>PROPOSITION</u> 1. Equivalently :

1. $\sigma \in F(R)$ is idempotent.
2. If $0 \to K \to M \to I \to 0$ is exact and if both K and I are σ-torsion modules, then M is a σ-torsion module.
3. Let $M,N \in M(R)$, N being σ-open in M, then the topology induced in N by the σ-topology in M coincides with the σ-topology in N.
4. If $B \subset A$ are left ideals of R such that $A \in T(\sigma)$ and if A/B is σ-torsion, then $B \in T(\sigma)$.

<u>Remark</u>. If $\sigma \in F(R)$ is idempotent, then $T(\sigma)$ is a multiplicatively closed set.

An $I \in M(R)$ is said to be <u>σ-injective</u> if, for every exact sequence $0 \to K \to M \to M/K \to 0$ with M/K being σ-torsion and any $f \in \mathrm{Hom}_R(K,I)$, there is an $\bar{f} \in \mathrm{Hom}_R(M,I)$ extending f to M.
A σ-injective module I is <u>faithfully</u> σ-injective if the extension $\bar{f}$ of f to M is unique.
It is easily verified that a necessary and sufficient condition for I to be faithfully σ-injective is that I is σ-torsion free.
If I is σ-injective then every $f \in \mathrm{Hom}_R(A,I)$ with $A \in T(\sigma)$, extends to

an $\bar{f} \in \mathrm{Hom}_R(R,I)$; this condition is clearly also a sufficient one.

Unless otherwise specified, σ will always be an idempotent kernel functor from now on.

If $M \in M(R)$ is σ-torsion free then there exists a faithfully σ-injective $I \in M(R)$, containing M, such that I/M is σ-torsion. This I is then unique up to isomorphism, it can be constructed as the extension of M, in some absolute injective hull E of M, maximal among submodules X of E with the property $\sigma(X/M) = X/M$. The faithfully σ-injective I containing M will be denoted by $Q_\sigma(M)$. The definition of $Q_\sigma(M)$ may be considered as the direct limit of the system

$$\{\mathrm{Hom}_R(A,M),\ \pi_{A,B} : \mathrm{Hom}_R(A,M) \to \mathrm{Hom}_R(B,M),\ A \supset B \in T(\sigma)\},$$

where $\pi_{AB}(f) = f|B$.

If M is not σ-torsion free then we put $Q_\sigma(M) = Q_\sigma(M/\sigma(M))$ and the direct limit interpretation yields at once that Q_σ is a covariant and left exact functor on $M(R)$, (if the R-module structure on the direct limit is defined in the usual way, cf. [12]). Moreover, $Q_\sigma(R)$ is a ring containing $R/\sigma(R)$ as a subring. The ring structure of $Q_\sigma(R)$ induces the R-module structure of $Q_\sigma(R)$ and it is unique as such. The ring $Q_\sigma(R)$ together with the canonical ring homomorphism $j : R \to Q_\sigma(R)$, provides us with a satisfying localization technique. In general, right exactness of Q_σ is not garanteed. Recall from [12] :

PROPOSITION 2. The following statements are equivalent :

1. Every $M \in M(Q_\sigma(R))$ is σ-torsion free.
2. For all $A \in T(\sigma)$, $Q_\sigma(R)j(A) = Q_\sigma(R)$.
3. Every $M \in M(Q_\sigma(R))$ is faithfully σ-injective.
4. For all $M \in M(R)$, $Q_\sigma(R) \otimes_R M \cong Q_\sigma(M)$.
5. The functor Q_σ is right exact and commutes with direct sums.

Let $\sigma \in F(R)$ be a, not necessarily idempotent, kernel functor. An R-module P is said to be σ-projective if : given σ-torsion free R-modules M, M' and an exact sequence $M' \to M \to 0$ together with an R-linear map $P \to M$, then there is a submodule P' of P such that P/P' is σ-torsion and there is an R-linear map $P' \to M'$ such that the diagram :

with rows exact, is commutative.
If P is absolute projective then it certainly is σ-projective. It is well-known, cf. [12], that Q_σ is right exact if and only if every $A \in T(\sigma)$ is σ-projective. However, if R is left Noetherian then Q_σ is right exact if and only if any one of the equivalent properties listed in Proposition 2 holds. An idempotent $\sigma \in F(R)$ having one of the properties mentioned in Proposition 2 is called a <u>T-functor</u>. In this case every left ideal of $Q_\sigma(R)$ is generated by a left ideal of $R/\sigma(R)$, so if R is left Noetherian then $Q_\sigma(R)$ is too.
Note that, for σ to be a T-functor it is sufficient that every $A \in T(\sigma)$ contains a $B \in T(\sigma)$ which is σ-projective because, from a morphism $h : A \to M$ we deduce the commutative diagram :

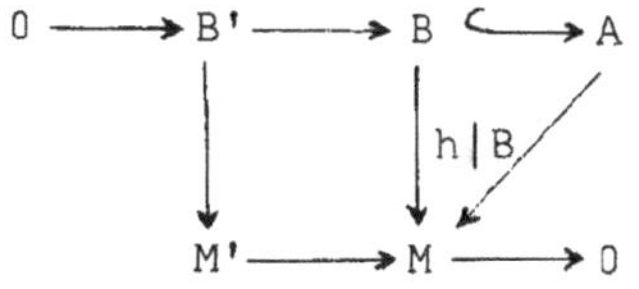

where A/B' is σ-torsion since $B' \in T(\sigma)$.
The second property in Proposition 2 states that every left ideal $A \in T(\sigma)$, becomes trivial under extension to the ring of quotients $Q_\sigma(R)$ for a T-functor σ.
L. Silver started investigating the correspondence between prime ideals

of R not in $T(\sigma)$ and prime ideals of $Q_\sigma(R)$. It has been shown that, in general, one cannot establish a one-to-one correspondence between these sets of ideals, (cf. [31], as a matter of fact this may be deduced from example 8 of [12], p. 47). The question is to determine a class of rings where the one-to-one correspondence does hold; we return to this problem in section II. In case R is a commutative ring, localization at a prime ideal P is in a natural way linked to the localization at a multiplicative system, i.e. $R - P$. In the noncommutative case however, several kernel functors may be associated to a prime ideal. We present a cross-cut of torsion theories used by A. Goldie, J. Lambek, G. Michler, A.G. Heinicke, D.C. Murdoch and the author.

I. 2. Localization at a Prime Ideal.

By definition, an ideal P of R is prime if and only if $R - P$ is an m-system, i.e., if $s_1, s_2 \in R - P$ then there is an $x \in R$ such that $s_1 x s_2 \in R - P$. If $R - P$ is a multiplicative set then P is said to be a completely prime ideal .

In the sequel, R will always be a left Noetherian ring, unless otherwise specified.

To a prime ideal P of R the multiplicative set $G(P) = \{g \in R,\ r \notin P \text{ implies } rg \notin P\}$, is associated. This set plays an important part in Goldie's localization theory, cf. [10], [11]. Define $\sigma_P \in F(R)$ by its filter $T(\sigma_P)$ consisting of left ideals A of R such that $[A : r] \cap G(P) \neq \phi$ for every $r \in R$. In [21], J. Lambek, G. Michler proved that σ_P coincides with the kernel functor determined by the injective hull $E(R/P)$ of $R/P \in M(R)$, ($M \in M(R)$ is torsion if $\mathrm{Hom}_R(M, E(R/P)) = 0$). O. Goldman defines a kernel functor τ_E by an injective module E as follows. For any $M \in M(R)$, put

$$\tau_E(M) = \cap \{\text{Kernels of } R\text{-linear maps } M \to E\}.$$

Hence, τ_E-torsion modules are given by $\mathrm{Hom}_R(M,E) = 0$ and this shows that for left Noetherian rings the kernel functor σ_P coincides with $\tau_{R/P} = \tau_{E(R/P)}$.

Another way of introducing localization at a prime ideal in a noncommutative ring is the following : to the m-system $R - P$, an idempotent kernel functor is associated such that its filter has a basis of ideals. A $\sigma \in F(R)$ is called a <u>bilateral functor</u> if and only if every $A \in T(\sigma)$ contains an ideal of R which is also in $T(\sigma)$. A bilateral functor σ is said to be <u>symmetric</u> if it is idempotent.

<u>PROPOSITION</u> 3. A bilateral $\sigma \in F(R)$ is symmetric if and only if $T(\sigma)$ is multiplicatively closed.

<u>PROOF</u>. Idempotency implies that $T(\sigma)$ is a multiplicative set. Conversely, if $A \in T(\sigma)$ and $B \subset A$ is such that $\sigma(A/B) = A/B$ then it is sufficient to prove that $B \in T(\sigma)$. Since R is left Noetherian, $A = Ra_1 + \ldots + Ra_n$ with $a_1,\ldots,a_n \in A$. Choose ideals $C_i \in T(\sigma)$ such that $C_i\, a_i \subset B$, which is possible because $\sigma(A/B) = A/B$. Putting $C = \bigcap_i C_i$ we obtain $C A \subset B$. Since $C,A \in T(\sigma)$ it follows that $C A \in T(\sigma)$ hence $B \in T(\sigma)$.

Transfer of the inclusion ordering for filters yields a partial ordering for kernel functors (see before). For a set $\{\sigma_\nu,\ \nu \in V\} \subset F(R)$ we define $\sigma = \inf_\nu \sigma_\nu$ by $\sigma(M) = \bigcap_\nu \sigma_\nu(M)$, for every $M \in M(R)$. Obviously $\sigma \leqslant \sigma_\nu$ for all $\nu \in V$ and if $\tau \leqslant \sigma_\nu$ for all $\nu \in V$ then $\tau \leqslant \sigma$. It is immediate from this definition that the inf of a set of idempotent kernel functors is idempotent and the inf of symmetric kernel functors is symmetric. Now let $\{\sigma_\nu,\ \nu \in V\}$ be a set of symmetric kernel functors and write $T(\nu)$ for the filter $T(\sigma_\nu)$. Let $T(\sigma)$ be the filter generated by all finite products of elements in $\bigcup_\nu T(\nu)$. This filter is associated to a symmetric functor σ with the property that, if $\rho \in F(R)$ is idempotent and $\rho \geqslant \sigma_\nu$ for all $\nu \in V$, then $\rho \geqslant \sigma$; this σ is the sup of $\{\sigma_\nu,\ \nu \in V\}$.

Now, to an m-system $R - P$ a symmetric kernel functor σ_{R-P} is associated. Let $M \in M(R)$ and define $\sigma_{R-P}(M) = \{m \in M,\ sRm = 0$ for some $s \in R-P\}$. The topology $T(R-P)$ corresponding to σ_{R-P} consists of the left ideals of R containing an ideal (s) generated by some $s \in R-P$. Observe that, in the absence of the left Noetherian condition for R, σ_{R-P} is still bilateral but not necessarily idempotent. In that case the idempotent closure $\bar{\sigma}_{R-P}$, in the sense of Goldman [12], may be defined by :

$$\bar{\sigma}_{R-P}(M) = \cap\ \{N,\ N \subset M,\ sRm \subset N,\ s \in R-P \text{ yields } m \in N\}.$$

Then $\bar{\sigma}_{R-P}$ is idempotent but not necessarily symmetric. Note also that $\bar{\sigma}_{R-P}(R)$ and $\sigma_{R-P}(R)$ are respectively the upper and lower $(R-P)$-component of the zero ideal as defined in [24]. Finally, to an arbitrary ideal A of R we may associate symmetric kernel functors σ_A and ${}_A\sigma$, i.e.,

$$\sigma_A = \inf\{\sigma_{R-P},\ P \supset A\},\quad {}_A\sigma = \sup\{\sigma_{R-P},\ P \supset A\}$$

and both depend only on the radical of A in R. Though both are plausible definitions, only the first will be used, (in section III). For an idempotent $\sigma \in F(R)$ define σ^0 as being the kernel functor corresponding to the filter based upon the ideals in $T(\sigma)$; σ^0 is the biggest symmetric kernel functor smaller than σ. Indeed, σ^0 is clearly bilateral and since $T(\sigma)$ is multiplicatively closed it follows immediately that σ^0 is symmetric. The relation between the Lambek-Michler torsion theory σ_P and the symmetric σ_{R-P} can be expressed as follows :

<u>PROPOSITION</u> 4. $\sigma_P^0 = \sigma_{R-P}$.

<u>PROOF</u>. If $A \in T(\sigma_P^0)$ then A contains an ideal $B \in T(\sigma_P)$, i.e. for every $r \in R$, $[B : r] \cap G(P) \neq \phi$. Since $B \subset [B : r]$ it follows that $B \in T(\sigma_P)$ is equivalent with $B \cap G(P) \neq \phi$, hence $B \in T(R-P)$ and also $A \in T(R-P)$, entailing $\sigma_P^0 \leqslant \sigma_{R-P}$. Moreover, if $\bar{x} \in \sigma_{R-P}(R/P)$ then $Cx \subset P$ for some

ideal $C \in T(R - P)$ and some $x \in \bar{x}$. Thus $sRx \subset P$ for some $s \in R - P$ and this yields $x \in P$ and $\bar{x} = 0$, implying that $\sigma_{R-P}(R/P) = 0$ which yields $\sigma_{R-P} \leq \tau_{R/P}$. Now, because $\tau_{R/P} = \sigma_P$, and σ_{R-P} is symmetric we get $\sigma_{R-P} \leq \sigma_P^0$.

Symmetric kernel functors have several advantages, f.i. these funtors play the main role in the construction of a structure sheaf on the prime-spectrum of a left Noetherian prime ring (section III). The price paid for the many advantages of using symmetric torsion theories is the difficulty of investigating property (T), due to the fact that principal left ideals generated by some $s \in R - P$ are not necessarily in $T(R - P)$. We say that a left ideal of R is <u>σ-closed</u>, or simply <u>closed</u>, if it is not in $T(\sigma)$. A left ideal A of R is <u>critical</u> if it is maximal among proper σ-closed left ideals for some idempotent $\sigma \in F(R)$. The set of critical left ideals for a fixed idempotent kernel functor σ will be denoted by $C'(\sigma)$.

<u>PROPOSITION</u> 5. A is a critical left ideal if and only if $A \in C'(\tau_{R/A})$.

<u>PROOF</u>. The "if" part is trivial. Suppose A is σ-critical. Consider the R-morphism $\pi : R \to R/A$ and let B be the R-submodule of R mapped onto $\sigma(R/A)$ under π. If $\sigma(R/A) \neq 0$ then B contains A properly and hence B is then in $T(\sigma)$ because $A \in C'(\sigma)$ for some idempotent functor σ. Therefore R/B is σ-torsion, but since $R/B \cong (R/A)/(B/A) = (R/A)/\sigma(R/A)$ it is also σ-torsion free, hence $R = B$ and $\sigma(R/A) = R/A$. The latter contradicts $A \in C'(\sigma)$, thus $\sigma(R/A) = 0$ and this yields that $\sigma \leq \tau_{R/A}$. Because $\tau_{R/A}(R/A) = 0$, we have that A is $\tau_{R/A}$-closed. Moreover, if B contains A properly while $B \notin T(\tau_{R/A})$ then $B \in T(\sigma) \subset T(\tau_{R/A})$, contradiction.

Let A be σ-critical and let s be an element not in A, then the R-linear map $\eta_s : R/[A : s] \to R/A$, defined by $x \bmod [A : s] \to xs \bmod A$, is a monomorphism. It follows that every submodule B of $R/[A : s]$ has the property $\sigma(R/B) = R/B$ and hence $[A : s]$ is a maximal left ideal not in $T(\sigma)$, or $[A : s] \in C'(\sigma)$. Recall also, cf. V. Dlab [8], that irreducible left ideals A and B of R are said to be <u>related</u> if there exist $s_1 \notin A$, $s_2 \notin B$ such that $[A : s_1] = [B : s_2]$ and that A and B are related if and only if the injectives $I(R/A)$ and $I(R/B)$ are isomorphic. Critical left ideals may also be considered as being the maximal elements in an equivalence class of related irreducible left ideals. This implies: a critical left ideal A is completely prime in its idealizer A_R in R, with $A_R = \{x \in R, Ax \subset A\}$. Indeed if $x \in R_A - A$ then $A \subset [A : x]$, and by maximality of A in the equivalence class we get $A = [A : x]$ and the fact that $A_R - A$ is a multiplicative set follows easily. Critical prime left ideals containing a prime ideal P of R are of particular interest.

<u>PROPOSITION</u> 6. Let P be a prime ideal of R. The following statements are equivalent :

1. A is maximal among left ideals not intersecting $G(P)$.
2. A is an irreducible prime left ideal of R containing P and $A \cap G(P) = \phi$.
3. A is a critical prime left ideal of R containing P and $A \cap G(P) = \phi$.

This and related properties may be found in [21]. The following problem may be connected to this. Let $\sigma = \sigma_{R - P}$ be the symmetric localization at the prime ideal P of R; characterize the σ-critical left ideals containing P. Proposition 6 fails to give a satisfying solution for this problem because the correspondence between $C'(\sigma_{R - P})$ and $C'(\tau_{R/P})$ is not known well enough.

The concept of a "critical left ideal" is strongly connected with Goldman's prime kernel functors, cf. [12]. Let $\sigma \in F(R)$ be idempotent.

A <u>support</u> for σ is a σ-torsion free R-module S such that, for every nonzero submodule S' of S we have that S/S' is σ-torsion. A $\sigma \in F(R)$ is called a <u>prime kernel functor</u> if there exists a support S for σ such that $\tau_S = \sigma$.
Clearly, if S is a support for σ, then any nonzero homomorphism from S to a σ-torsion free R-module has to be injective. If σ is prime and if S is any support for σ then $\tau_S = \sigma$. Moreover, there exists a unique (up to isomorphism) support for σ which at the same time is σ-injective.

<u>PROPOSITION</u> 7. Let $\sigma \in F(R)$ be idempotent and let A be a σ-critical left ideal in R, then :

1. A is $\tau_{R/A}$-critical.
2. The quotient module R/A is a support for σ.
3. The induced kernel functor $\tau_{R/A}$ is prime.

<u>PROOF</u>. The first statement follows from proposition 5. Secondly, every nonzero submodule of R/A is image of some left ideal of R which contains A properly and as such, it is σ-open and thus R/A is a support for σ. The last statement follows immediately from 1 and 2.

<u>PROPOSITION</u> 8. A $\sigma \in F(R)$ is idempotent if and only if $\sigma = \inf\{\tau_{R/A}, A \in C'(\sigma)\}$.

<u>PROOF</u>. The fact that $\sigma(R/A) = 0$ for any $A \in C'(\sigma)$ implies that $\sigma \leq \tau_{R/A}$ for every $A \in C'(\sigma)$ and thus $\sigma \leq \inf\{\tau_{R/A}, A \in C'(\sigma)\}$. Let $\tau \geq \sigma$ with $\tau \neq \sigma$. Then there is a $C \in T(\tau) - T(\sigma)$. Since $C \notin T(\sigma)$ we may find an $A_1 \in C'(\sigma)$ such that $C \subset A_1$. For this particular A_1 it follows that $\tau \leq \tau_{R/A_1}$ cannot hold, hence $\sigma = \inf\{\tau_{R/A}, A \in C'(\sigma)\}$. To prove the converse, define $C'(\sigma)$ to be the set of maximal σ-closed left ideals of R maximal in an equivalence class of related irreducible left ideals.

Since $\tau_{R/A}$ is idempotent for any $A \in C'(\sigma)$, σ is idempotent too.

COROLLARIES. If $\sigma \in F(R)$ is idempotent then $\sigma = \tau_M$ where M is the direct sum of the non-isomorphic quotient modules R/A for all $A \in C'(\sigma)$. Furthermore, $\sigma = \tau_N$ where $N = \amalg Q_\sigma(R/A)$, the direct sum ranging over all $A \in C'(\sigma)$. It is clear that M (or N) cannot be a support for σ if there exist at least two factors in the sum, whence the following results. An idempotent $\sigma \in F(R)$ is a prime kernel functor if and only if $Q_\sigma(R/A) \cong E$ for all $A \in C'(\sigma)$.

Alternative ways of looking at critical left ideals are encountered in [19], [32]; they may be described as left annihilators of the elements of indecomposable injective modules, so they are related to what is called an atom in [32].

For completeness sake, let us recall that the left Artinian condition for R is equivalent to every critical prime left ideal being a maximal left ideal of R. Artinian conditions will be avoided in the present context.

The correspondence between prime ideals of R and prime ideals of $Q_\sigma(R)$ has been studied in case $\sigma = \sigma_P$ in [21], [13], [31]. In order to get useful results one has to impose the left Ore condition on R.
Let P be a prime ideal of R, with associated multiplicative set $G(P)$ as before. Then R is said to satisfy the left Ore condition with respect to $G(P)$ if for any $x \in R$, $g \in G(P)$, there exist $x' \in R$ and $g' \in G(P)$ such that $g'x = x'g$. The image of P under $R \to R/\sigma_P(R)$ will be denoted by $\bar{P}$. Then $G(\bar{P}) = (G(P) + \sigma_P(R))/\sigma_P(R)$, and by straightforward argumentation one derives that R satisfies the left Ore condition with respect to $G(P)$ if and only if $R/\sigma_P(R)$ satisfies the left Ore condition with respect to $G(\bar{P})$. From [21] Proposition 5.5., it follows that R satisfies the left Ore condition with respect to $G(P)$ if and only if the elements of $G(\bar{P})$ are units in $Q_{\sigma_P}(R)$. This is also equivalent to $Q_{\sigma_P}(P)$ being the Jacobson radical of $Q_{\sigma_P}(R)$; and $Q_{\sigma_P}(R/P)$ is then isomorphic to the

classical ring of quotients $Q_{cl}(R/P)$.
Moreover, σ_P has property (T) and $Q_{\sigma_P}(R)$ is a simple Artinian ring. The aim of the following section is to derive more or less similar results in case R is a left Noetherian (prime) ring, with respect to localization at symmetric T-functors.

Special references for Section I.

V. DLAB [8]; P. GABRIEL [9]; A.W. GOLDIE [10], [11]; O. GOLDMAN [12]; A.G. HEINICKE [13]; J. LAMBEK [19], [20]; J. LAMBEK, G. MICHLER [21]; D.C. MURDOCH [24]; D.C. MURDOCH, F. VAN OYSTAEYEN [26], [27]; S.K. SIM [30], [31]; H. STORRER [32].

II. SYMMETRIC KERNEL FUNCTORS

II. 1. Localization at Symmetric Kernel Functors.

Unless otherwise specified, R is left Noetherian and σ is a symmetric T-functor. The canonical R-module morphism $j : R \to Q_\sigma(R)$ is a ring homomorphism. For a left ideal A of R, the extension $Q_\sigma(R)j(A)$ of $j(A)$ to a left ideal of $Q_\sigma(R)$ will be denoted by A^e. On the other hand, if B is a left ideal of $Q_\sigma(R)$ then $B^c = j^{-1}(B)$ is said to be the contraction of B to R.

THEOREM 9. For every left ideal B of $Q_\sigma(R)$, $B^{ce} = B$. For every left ideal A of R, $A^{ec} = A_\sigma$ where $A_\sigma = \{x \in R, Cx \subset A$ for some $C \in T(\sigma)\}$.

PROOF. To prove the first assertion, let $b \in B$. Since $Q_\sigma(R)/j(R)$ is σ-torsion, there is a $C \in T(\sigma)$ such that $Cb \subset j(R)$. By property (T) we have that $Q_\sigma(R)j(C) = Q_\sigma(R)$ (see Proposition 2), hence from $Cb = j(C)b$ it follows that $Q_\sigma(R)Cb = Q_\sigma(R)b$ or $b \in Q_\sigma(R)(B \cap j(R)) = B^{ce}$, entailing $B = B^{ce}$. To prove the second statement note first that $A^{ec} = j^{-1}(Q_\sigma(R)j(A) \cap j(R))$. If $x \in A_\sigma$ then $Cx \subset A$ for some ideal $C \in T(\sigma)$. Hence $Q_\sigma(R)j(C)j(x) \subset Q_\sigma(R)j(A)$ and then property (T) yields $j(x) \in A^e$ and $x \in A^{ec}$. Conversely let $x \in A^{ec}$, i.e., $j(x) \in Q_\sigma(R)j(A) \cap j(R)$. Thus we may write $j(x) = \Sigma' q_i a_i$ with $q_i \in Q_\sigma(R)$, $a_i \in j(A)$. Now choose an ideal C in $T(\sigma)$ such that $Cq_i \subset j(R)$ for all i. Then $Cj(x) \subset j(A)$ and $Cx \subset A + \sigma(R)$. By the left Noetherian property for R we can find an ideal C' in $T(\sigma)$, such that $C'\sigma(R) = 0$ and hence $C'Cx \subset A$. Idempotency of σ implies that $C'C \in T(\sigma)$ and $x \in A_\sigma$.

COROLLARY 1. Let I be a left ideal of R. It is easily verified that $Q_\sigma(I) = Q_\sigma(R)j(I)$.

COROLLARY 2. There is one-to-one correspondence between maximal left ideals of $Q_\sigma(R)$ and elements of $C'(\sigma)$.

Proof of the last statement : Suppose that $A \in C'(\sigma)$. It is obvious that A^e is a maximal left ideal of $Q_\sigma(R)$ since a proper left ideal M, containing A properly, restricts to A and thus $M^{ce} = A^e$ would contradicht $M \neq A^e$. Conversely, if M is a maximal left ideal of $Q_\sigma(R)$ then M^c is proper in R and $M^c \notin T(\sigma)$ by property (T). But then $M^c \subset A$ for some $A \in C'(\sigma)$ and therefore $M^{ce} \subset A^e$ or $M \subset A^e$ follows. Thus $M = A^e$.

If A is an ideal of R then A_σ is an ideal of R but A^e is not necessarily an ideal of $Q_\sigma(R)$. Consequently an exact sequence of ring homormorphisms $0 \to K \to R \xrightarrow{\pi} R/K \to 0$ does not always yield a ring homomorphism $Q_\sigma\pi : Q_\sigma(R) \to Q_\sigma(R/K)$, even if Q_σ is exact. This problem is central in what follows, most of the following results apply to section III.

THEOREM 10. Let $\tau \geq \sigma$ be arbitrary symmetric kernel functors, then :

1. $Q_\sigma(\tau(R)) = \tau(Q_\sigma(R))$ and $Q_\tau(Q_\sigma(R)) \cong Q_\tau(R)$.
2. The unique R-linear map $Q_\sigma(R) \to Q_\tau(R)$ extending the canonical $j_\tau : R \to Q_\tau(R)$ to $Q_\sigma(R)$, is a ring homomorphism for the ring structure induced in $Q_\sigma(R)$ and $Q_\tau(R)$ by their respective R-module structure.

PROOF 1. The exact sequence $0 \to \tau(R) \to R \to R/\tau(R) \to 0$ yields under localization an exact sequence :

$$0 \to Q_\sigma(\tau(R)) \to Q_\sigma(R) \to Q_\sigma(R/\tau(R))$$

If we are able to proof that $\tau(Q_\sigma(R/\tau(R))) = 0$ then $\tau(Q_\sigma(R)) \subset Q_\sigma(\tau(R))$ follows, and equality is immediate. Pick an $x \in \tau(Q_\sigma(R/\tau(R)))$. Then $Bx = 0$ for some $B \in T(\sigma)$ while $Ax \subset R/\tau(R)$ for some $A \in T(\sigma)$. Since A,B may be chosen to be ideals of R we get $BA \subset B$, hence $BA\,x = 0$. This entails that $Ax \subset \tau(R/\tau(R)) = 0$ and thus $x = 0$. Moreover

$R/\sigma(R) \cap \tau(Q_\sigma(R)) = \tau(R)/\sigma(R)$ yields inclusions :

$$R/\tau(R) \hookrightarrow Q_\sigma(R)/\tau(Q_\sigma(R)) \hookrightarrow Q_\tau(R)$$

and therefore $Q_\tau(Q_\sigma(R)) \cong Q_\tau(R)$.

2. Since $\tau \geqslant \sigma$ and $\sigma(Q_\tau(R)) = 0$ it follows that $Q_\tau(R)$ is faithfully σ-injective and so the R-module structure of $Q_\tau(R)$ extends uniquely to give $Q_\tau(R)$ a $Q_\sigma(R)$-module structure, which by the uniqueness of it, should coincide with the structure induced by ring multiplication in $Q_\tau(R)$. Let J_τ be the unique R-linear map $Q_\sigma(R) \to Q_\tau(R)$ extending j_τ and let ξ,η be elements of $Q_\sigma(R)$.
We may find a $C \in T(\sigma)$ such that $C\xi \subset R/\sigma(R)$. Then $J_\tau(C\xi\eta) = C\,J_\tau(\xi\eta)$, but also $J_\tau(C\xi\eta) = C\xi.J_\tau(\eta) = C.\xi J_\tau(\eta)$, by definition of the $Q_\sigma(R)$-module structure. We derive form this that $J_\tau(\xi\eta) - \xi J_\tau(\eta) \in \sigma(Q_\tau(R)) = 0$, in other words, J_τ is $Q_\sigma(R)$ linear. We are left to prove that $Q_\sigma(\tau(R))$ is an ideal of $Q_\sigma(R)$. By 1. it is obviously a right ideal and a left R-module. If $x \in \tau(Q_\sigma(R))$ and $\lambda \in Q_\sigma(R)$ then $Bx = 0$ for some ideal $B \in T(\tau)$, while $A\lambda \subset R/\sigma(R)$ for some $A \in T(\sigma)$.
Finally, $BA\lambda\, x = 0$ with $BA \in T(\tau)$ entails that $\lambda x \in \tau(Q_\sigma(R))$.

<u>DEFINITION</u>. Let σ be a symmetric T-functor, R a left Noetherian ring. R is said to be <u>σ-perfect</u> if every proper σ-closed ideal A of R extends to a proper ideal A^e of $Q_\sigma(R)$.

<u>PROPOSITION</u> 11. Suppose that R is a σ-perfect ring and let P be a σ-closed prime ideal of R. Then P^e is a prime ideal of $Q_\sigma(R)$.

<u>PROOF</u>. By the left Noetherian property for R, we have that $CP_\sigma \subset P$ for some $C \in T(\sigma)$. Hence, since C is not contained in P, $P = P_\sigma$ follows. Now P^e is an ideal of $Q_\sigma(R)$ because we assumed R to be σ-perfect. Suppose $AB \subset P^e$ for some left ideals A,B of $Q_\sigma(R)$. Then we have that $A^cB^c \subset (AB)^c \subset P^{ec} = P$ and therefore A^c or B^c is contained in P,

yielding that $A^{ce} = A$ or $B^{ce} = B$ is contained in P^e.

COROLLARY. With the above assumptions : there is a one-to-one correspondence between proper prime ideals of $Q_\sigma(R)$ and prime ideals of R which are σ-closed. This is easily seen by verifying that proper prime ideals P of $Q_\sigma(R)$ restrict to σ-closed prime ideals of R. Indeed, if A, B are ideals of R such that $AB \subset P^c$ then $(AB)^e \subset P^{ce} = P$. Consequently $A \in T(\sigma)$ yields $B^e \subset P$ and $B \subset P^c$ while $A \notin T(\sigma)$ yields $A^eB^e \subset P$, thus A^e or B^e is contained in P entailing that A or B is in P^c.

PROPOSITION 12. Let R be a σ-perfect ring, then :

1. For every ideal A of R, $\text{rad } A^e = (\text{rad } A)^e$.
2. There is a one-to-one correspondence between σ-closed left P-primary ideals of R and left P^e-primary ideals of $Q_\sigma(R)$.

PROOF. The previous proposition yields that $\text{rad } A^e$ is intersection of the extended ideals P^e with $P^e \supset A^e$. Hence

$$(\text{rad } A^e)^c = \cap \{P,\ P \supset A \text{ and } P \notin T(\sigma)\}.$$

If $(\text{rad } A^e)^c \subset (\text{rad } A)_\sigma$ then $(\text{rad } A)^e = (\text{rad } A^e)^{ce} = \text{rad } A^e$ will follow. Therefore, take $x \in P$ for all $P \supset A$ such that P is σ-closed. Let P_0 be an arbitrary prime ideal in $T(\sigma)$, such that $P_0 \supset A$. Then, if $x \notin P_0$, there is an ideal $C_0 \in T(\sigma)$ for which $C_0x \subset P_0$. Because the number of minimal prime ideals containing A is finite, there exists an ideal $C \in T(\sigma)$ for which $Cx \subset P$ for every minimal prime P containing A such that $P \in T(\sigma)$. However, since x (and therefore certainly Cx) is contained in all σ-closed minimal prime ideals containing A, it follows that $Cx \subset \text{rad } A$ and $x \in (\text{rad } A)_\sigma$.

2. Recall that an ideal I of R is said to be left primary if $AB \subset I$ implies $B \subset I$ or $A \subset \text{rad } I$. Since R is left Noetherian it follows that

rad I is a prime P, and I is called a left P-primary ideal. Again, $CI_\sigma \subset I$ for some ideal $C \in T(\sigma)$. Then $P \not\supset C$ forces $I_\sigma = I$ and using 1. the proof becomes easy, following the lines of the proof of Proposition 11.

Remark. If R is left Noetherian, and σ being idempotent, then every σ-closed ideal of R is contained in a maximal σ-closed ideal. Let P be a maximal element in the set of σ-closed ideals, then P is a prime ideal. For, let A and B be ideals of R such that $AB \subset P$ with $A \not\subset P$ and $B \not\subset P$, then we have that $A + P$ and $B + P$ are in $T(\sigma)$. Hence $(A + P)(B + P) \subset P$ contradicts $P \notin T(\sigma)$. The set $C(\sigma) = \{P$, P maximal in the set of σ-closed ideals$\}$ determines σ completely in case σ is symmetric, then $T(\sigma)$ is the set of left ideals of R containing an ideal which is not contained in any element of $C(\sigma)$.

LEMMA 13. Let R be an arbitrary ring and let σ be a T-functor on $M(R)$. If P is a left ideal of R then $Q_\sigma(P)$ is maximal in the set of R-submodules X in $Q_\sigma(R)$ containing $P/\sigma(P)$ such that $X/(P/\sigma(P))$ is a σ-torsion module.

PROOF. Denote $P/\sigma(P)$ by $\overline{P}$. Property (T) implies that

$$Q_\sigma(Q_\sigma(R)/\overline{P}) = Q_\sigma(R)/Q_\sigma(P) = Q_\sigma(R/P),$$

and we may derive the following exact sequence :

$$0 \to Q_\sigma(P)/\overline{P} \to Q_\sigma(R)/\overline{P} \to Q_\sigma(Q_\sigma(R)/\overline{P}) \to 0.$$

The R-module $Q_\sigma(Q_\sigma(R)/\overline{P})$ is σ-torsion free, thus $\sigma(Q_\sigma(R)/\overline{P}) \subset Q_\sigma(P)/\overline{P}$, but since $Q_\sigma(P)/\overline{P}$ is σ-torsion equality follows. Obviously $Q_\sigma(P)$ is maximal with the desired property.

We return to the case where R is a left Noetherian prime ring, and σ is a symmetric T-functor.

DEFINITION. An ideal A of R is a σ-ideal if for all $C \in T(\sigma)$ we have that $[AC : A] \in T(\sigma)$, i.e., for a $C \in T(\sigma)$ there exists a $C' \in T(\sigma)$ such that $C'A \subset AC$.

THEOREM 14. Let R be left Noetherian and prime, let σ be a symmetric T-functor. For an ideal A of R, the following statements are equivalent :

1. A is a σ-ideal.
2. A^e is an ideal of $Q_\sigma(R)$.

PROOF. Because σ is a T-functor the extended ideal A^e is $Q_\sigma(A)$ and $A^e \cap R = A_\sigma$. Consider $A^e Q_\sigma(R)$. If $x \in A^e Q_\sigma(R)$ then we may write $x = \Sigma' q_i a_i q_i'$ with $a_i \in A$ and $q_i, q_i' \in Q_\sigma(R)$ and the sum being finite. Therefore, $C'q_i \subset R$ for all i, for a well-chosen $C' \in T(\sigma)$. Moreover, there is a $C'' \in T(\sigma)$ such that $C''a_i \subset AC'$, by the σ-ideal condition for A.
Finally, there is a $C \in T(\sigma)$ such that $Cq_i \subset C''$ for all i, thus $Cx \subset A$. The foregoing lemma then states that $x \in Q_\sigma(A)$ and that finishes the proof of the fact that 1. implies 2..
Conversely, suppose that A^e is an ideal of $Q_\sigma(R)$. Then

$$(AC)^e = Q_\sigma(R)AC = A^e C = A^e Q_\sigma(R) C = A^e,$$

thus A^e/AC is σ-torsion, à fortiori A/AC is σ-torsion entailing that there is a $C' \in T(\sigma)$ such that $C'A \subset AC$ and thus A is a σ-ideal.
Note that the implication, $1 \Rightarrow 2$, is obviously true for arbitrary T-functors.

COROLLARY. In case every ideal $C \in T(\sigma)$ contains a central element generating an ideal in $T(\sigma)$, then obviously, R is σ-perfect. Therefore, matrix rings over commutative prime rings are σ-perfect for any T-functor σ.

The σ-ideal condition, (written element-wise), resembles the left Ore condition. We investigate the correlation between these conditions in case $\sigma = \sigma_P$ for some prime ideal P of R. The left Ore condition with respect to $G(P)$ is equivalent with $Rs \in T(\sigma_P)$ for all $s \in G(P)$. Indeed, if $Rs \in T(\sigma_P)$ then for all $r \in R$, $[Rs : r] \cap G(P) \neq \phi$, i.e., there exists an $s' \in G(P)$ such that $s'r \in Rs$, or, for every $r \in R$, $s \in G(P)$ there exists s', r' such that $s'r = r's$. Conversely the left Ore condition obviously implies that for arbitrary $s \in G(P)$ and $r \in R$ there exists an $s' \in [Rs : r]$. The extra assumption that R satisfies the left Ore condition yields that condition 1 of Proposition 14 is equivalent to the (P,A)-condition, i.e., for every $s \in G(P)$ there exists an $s' \in G(P)$ such that $s'A \subset As$. Summarizing this, we have proved

PROPOSITION 15. Let R be a left Noetherian prime ring satisfying the left Ore condition with respect to $G(P)$ for some prime ideal P of R. Then, if A satisfies the (P,A)-condition, A^e is an ideal of $Q_{\sigma_P}(R)$. If σ_P is symmetric then the converse is also true.

PROPOSITION 16. Let R be a left Noetherian prime ring. Let τ be an idempotent kernel functor such that $\sigma = \tau^0$ is a T-functor. If $P \in C(\tau^0)$ is such that $Q_\tau(P)$ is a proper ideal of $Q_\tau(R)$ then $P^e = Q_\sigma(R)P$ is a proper ideal of $Q_\sigma(R)$.

PROOF. Since $\tau \geq \sigma$, we have an inclusion $Q_\sigma(R) \hookrightarrow Q_\tau(R)$. Then

$$P^e \subset Q_\tau(R)P \subset Q_\tau(P), \; I = Q_\tau(P) \cap Q_\sigma(R)$$

is an ideal containing P^e and therefore I restricts to P or R, hence

$I = Q_\sigma(R)$ or $I^c = P$. But as $1 \notin Q_\tau(P)$ it follows that $I = P^e$.

COROLLARY. If σ_{R-P} is a T-functor, while R satisfies the left Ore condition with respect ot G(P), then P^e is an ideal of Q_{R-P} and P^e is the intersection of $Q_\sigma(R)$ with the Jacobson radical of $Q_{\sigma_P}(R)$. One easily checks that if σ_{R-P} is a T-functor and if the elements of G(P) are units in $Q_{R-P}(R)$ then $Rs \in T(\sigma_{R-P})$ for all $s \in G(P)$, also we have that $\sigma_P = \sigma_{R-P}$ and R satisfies the left Ore condition.

PROPOSITION 17. Let R be a left Noetherian prime ring satisfying the left Ore condition with respect to G(P) for some prime ideal P of R. Suppose that $\sigma_P = \overset{0}{\sigma}_P (= \sigma_{R-P})$. Then :

1. The (P,P)-condition holds.
2. The Jacobson radical of $Q_{\sigma_P}(R)$ is equal to P^e and $Q_{\sigma_P}(R/P)$ is a simple Artinian ring.
3. $P = \cap \{A, A \in C'(\sigma)\}$.

PROOF. P is known to extend to an ideal under localization at σ_P, cf. [21]. It has been noted that this yields the σ_P-ideal condition and that condition 1 of Proposition 14 transforms to the (P,P)-condition under the additional hypothesis that R satisfies the left Ore condition.

2. This is a consequence of [21]; see the remarks on p. 14 and p. 15.

3. From Corollary 2 to Theorem 9 we derive that M is a maximal left ideal of $Q_{\sigma_P}(R)$ if and only if $M^c \in C'(\sigma_P)$. Now, because of the second statement, we may write $P^e = \cap \{A^e, A \in C'(\sigma_P)\}$ and thus, by contraction, $P = (\cap A^e)^c = \cap A^{ec} = \cap A$.

Property 3 above, will reappear in the next section.

II. 2. Quasi-prime Kernel Functors.

In this section quasi-prime kernel functors are defined in such a way that each σ_{R-P} is quasi-prime. This ameliorates the partial correspondence between prime kernel functors and certain prime ideals which is implicite in Goldman's set up. Throughout this section R will be a left Noetherian ring and σ will be a symmetric kernel functor. For any subset S of R let $[A : S]$ denote the left ideal $\{x \in R, xS \subset A\}$. So $[A : R]$ will be the biggest ideal contained in A. If B is a left ideal then $[A : B]$ is an ideal. Since σ is symmetric and because $[A : S]$ contains every ideal contained in A it follows that $[A : S]$ is in $T(\sigma)$ whenever A is. If S,T are subsets of R then $[A : TS] = [[A : S] : T]$. A symmetric kernel functor is called a restricted kernel functor if $A \in C'(\sigma)$ implies that $[A : R] \in C(\sigma)$.

PROPOSITION 18. 1. A symmetric kernel functor is restricted if and only if for each $A \in C'(\sigma)$ there is a $P \in C(\sigma)$ such that $[A : R] \subset P$ and $[A + P : R] = P$.

2. If for every $P \in C(\sigma)$ and every $A \in C'(\sigma)$ there can be found an ideal $I \in T(\sigma)$ such that $[A : I] = P$ then $C(\sigma) = \{P\}$ and $\sigma = \sigma_{R-P}$.

PROOF. 1. If σ is restricted then $P = [A : R]$ has the desired property. Conversely, let $A \in C'(\sigma)$ with $[A : R] \subset P$ and $[A + P : R] = P \in C(\sigma)$. Hence $A + P \notin T(\sigma)$ since otherwise P would be in $T(\sigma)$ too and this contradicts $P \in C(\sigma)$. Now $A \in C'(\sigma)$ yields $P \subset A$ and $P = [A : R]$ follows.

2. Let $P,P' \in C(\sigma)$ and choose $A \in C'(\sigma)$ such that $P \subset A$, i.e., $P = [A : R]$. If I is an ideal such that $P_1 = [A : I] \supset [A : R] = P$ then $P = P_1$ follows because P and P_1 are both elements of $C(\sigma)$. Hence $C(\sigma) = \{P\}$ and $\sigma = \sigma_{R-P}$ follows.

COROLLARY. σ_{R-P} is restricted if and only if $P = \cap \{A, A \in C'(\sigma)\}$.

Next proposition generalizes part of Proposition 17.

PROPOSITION 19. Let σ stand for σ_{R-P} and suppose that σ is a T-functor, then the following statements are equivalent :

1. σ is a restricted kernel functor.
2. The extension P^e of P to $Q_\sigma(R)$ is the Jacobson radical of $Q_\sigma(R)$.

PROOF. The Jacobson radical $J(Q_\sigma(R))$ is the intersection of the maximal left ideals of $Q_\sigma(R)$, hence by the corollary to Theorem 9 we get :

$$J(Q_\sigma(R)) = \cap \{A^e, A \in C'(\sigma)\}.$$

Thus, if $P^e = \cap A^e$ then $P = P^{ec} = (\cap A^e)^c = \cap A$ and this implies that σ is restricted. Conversely if $P = \cap \{A, A \in C'(\sigma)\}$ then $P^e = (\cap A)^e \subset \cap A^e$ and by contraction $P = P^{ec} \subset (\cap A^e)^c = \cap A^{ec} = \cap A = P$. Hence, $P^e(\cap A^e)^{ce} = \cap A^e = J(Q_\sigma(R))$.

The following gives sufficient conditions for an element A in $C'(\sigma)$ to be a prime left ideal.

PROPOSITION 20. 1. If $A \in C'(\sigma)$ is such that $[A : R] \in C(\sigma)$ then A is a prime left ideal of R.

2. Let σ be a symmetric T-functor and suppose that R is a σ-perfect ring. In this case, every left ideal $A \in C'(\sigma)$ is a prime left ideal.

PROOF. 1. Suppose there exist left ideals B, C of R such that $B \not\subset A$ and $C \not\subset A$ but $BC \subset A$. Therefore $(BR + [A : R])(C + A) \subset A$, but since $C + A$ and $BR + [A : R]$ are in $T(\sigma)$ and because $T(\sigma)$ is multiplicatively closed it follows that $A \in T(\sigma)$ contradictory to the hypothesis that $A \in C'(\sigma)$.

2. Suppose we have left ideals B,C such that $BC \subset A$, then $(BR)C \subset A$. Now $(BR)^e = Q_\sigma(R)(BR)$ is an ideal of $Q_\sigma(R)$ and this entails $(BR)^e C \subset A^e$. Theorem 9, Corollary 2, yields that A^e is a maximal left ideal of $Q_\sigma(R)$ and thus $C^e \not\subset A^e$ yields $C^e + A^e = Q_\sigma(R)$ and we derive from

$$(BR)^e = (BR)^e C^e + (BR)^e A^e \subset A^e,$$

by contraction, that either $(BR)_\sigma$ or C_σ is in $A_\sigma = A$.

COROLLARY 1. If σ is a restricted kernel functor then every $A \in C'(\sigma)$ is a prime left ideal.

COROLLARY 2. If σ_P is symmetric and restricted then the elements of $C'(\sigma_P)$ are exactly the left ideals of R maximal in the set of ideals not intersecting $G(P)$; see Proposition 6.

An R-module M is an R-bimodule if M is a right R-module such that : $x(my) = (xm)y$ for all $m \in M$ and $x,y \in R$.

DEFINITION. Let σ be an arbitrary idempotent kernel functor. An R-bimodule S is said to be a quasi-support for σ if :

1. S is a σ-torsion free R-module.
2. For every nonzero sub-bimodule $S' \subset S$, the quotient S/S' is a σ-torsion R-module.

PROPOSITION 21. If S is a quasi-support for an idempotent kernel functor σ, then :

1. S is an essential extension of every nonzero sub-bimodule of S.
2. Every nonzero sub-bimodule of S is a quasi-support for σ.
3. If T is a bimodule such that $T \supset S$ with $\sigma(T) = 0$ and $\sigma(T/S) = T/S$, then T is also a quasi-support for σ.

4. If τ_S is the kernel functor associated with S, then S is a quasi-support for τ_S.
5. If S contains a sub-bimodule S' which is a support for σ then S is a support for σ.

The proofs of these assertions follow the same lines as the proofs of the corresponding properties for supporting modules, cf. [12].

PROPOSITION 22. Let $\sigma \in F(R)$ be idempotent and let P be an ideal in R. Then R/P is a quasi-support for σ if and only if $P \in C(\sigma)$.

PROOF. Suppose that R/P is a quasi-support, then every ideal I of containing P properly, gives rise to a σ-torsion module R/I because I/P is a sub-bimodule of R/P. Hence, $I \in T(\sigma)$ for every I properly containing P. Therefore, $\sigma(R/P) = 0$ entails $P \in C(\sigma)$. Conversely, from $P \in C(\sigma)$ and $\sigma(R/P) = 0$ we may derive that for every ideal I properly containing P the quotient module R/I is σ-torsion. Hence, any sub-bimodule M of R/P has the property that (R/P)/M is σ-torsion; thus, R/P is a quasi-support for σ.

DEFINITION. A symmetric kernel functor σ is said to be a quasi-prime if there exists a quasi-support S for σ such that $\sigma = \overset{0}{\tau}_S$.
Proposition 4 together with the foregoing, yield that every σ_{R-P} associated with a prime ideal P of R is a quasi-prime.

Remark. If R is a left Noetherian ring such that for each prime ideal P of R, every essential left ideal of R/P contains a nonzero ideal, then there is a one-to-one correspondence between idecomposable injective R-modules I and prime ideals of R. Let I_P be the injective module corresponding to the prime ideal P. The torsion theory induced by I_P in M(R) coincides with σ_P (cf. [21]). Since $\sigma_{R-P} = \overset{0}{\sigma}_P$, we also get a one-to-one correspondence between ideals P of R and quasi-prime kernel functors associated with an indecomposable injective module.

PROPOSITION 23. If the elements of $\{Q_\sigma(R/P),\ P \in C(\sigma)\}$ are isomorphic to one another then σ is a quasi-prime.

PROOF. Let $P \in C(\sigma)$, i.e., R/P is a quasi-support for σ. Suppose that σ' is a symmetric kernel functor different from σ and $\sigma' \geq \sigma$. An ideal $A \in T(\sigma') - T(\sigma)$ is contained in some element P of $C(\sigma)$, while R/P is σ'-torsion. Moreover, since $Q_\sigma(R/P)/(R/P)$ is σ-torsion it is certainly σ'-torsion. Therefore, the exact sequence :

$$0 \to R/P \to Q_\sigma(R/P) \to Q_\sigma(R/P)/(R/P) \to 0$$

yields that $Q_\sigma(R/P)$ is σ'-torsion for the $P \in C(\sigma)$ with $P \supset A$, hence for every $P \in C(\sigma)$ by the hypothesis. Hence, σ is the largest symmetric kernel functor for which $E \cong Q_\sigma(R/P)$ is torsion free, i.e., $\sigma = \tau^0_E$. The fact that $Q_\sigma(R/P)/(R/P)$ is σ-torsion yields that $\sigma = \tau^0_{R/P}$, concluding the proof.

PROPOSITION 24. A symmetric $\sigma \in F(R)$ is equal to $\inf \tau^0_{R/P} = (\inf \tau_{R/P})^0$, the inf being taken over the elements $P \in C(\sigma)$.

PROOF. As before if $\sigma' \geq \sigma$ and σ' different from σ then for some $P \in C(\sigma)$ we have that $\sigma'(R/P) = R/P$ and thus σ' cannot be smaller than $\tau^0_{R/P}$. On the other hand, since $\sigma(R/P) = 0$, $\sigma \leq \tau^0_{R/P}$ follows.

COROLLARY 1. If σ, σ' are symmetric kernel functors such that $\sigma' \geq \sigma$ and $\sigma'(R/P) \neq R/P$ for all $P \in C(\sigma)$, then $\sigma = \sigma'$.

COROLLARY 2. Let S be a quasi-support for σ, then for every $\sigma' \geq \sigma$, σ' different from σ such that $\sigma'(S) \neq 0$ we have that $\sigma'(S) = S$.

PROOF. $\sigma'(S)$ is a sub-bimodule of S, hence $S/\sigma'(S)$ is σ-torsion à fortiori σ'-torsion, thus $S = \sigma'(S)$.

COROLLARY 3. If for all $P \in C(\sigma)$, the induced symmetric kernel functors $\overset{0}{\tau}_{R/P}$ coincide, then : $\sigma = \overset{0}{\tau}_{R/P}$, each R/P is a quasi-support for σ and σ is quasi-prime.

PROPOSITION 25. Let σ be a symmetric kernel functor. If $A \in C'(\sigma)$ is such that $[A : R] = P$ is a prime ideal then $\overset{0}{\tau}_{R/A} = \overset{0}{\tau}_{R/P}$.

PROOF. Let $0 \neq \bar{x} \in R/P$ and suppose there is an ideal $I \in T(\overset{0}{\tau}_{R/A})$ such that $I\bar{x} = 0$. Then, $I(x) \subset P$ for some $x \notin P$ representing $\bar{x}$, hence $I \subset P \subset A$ and this yields $A \in T(\overset{0}{\tau}_{R/A}) \subset T(\tau_{R/A})$, contradiction. Thus, R/P is $\overset{0}{\tau}_{R/A}$-torsion free and $\overset{0}{\tau}_{R/A} \leqslant \overset{0}{\tau}_{R/P}$. Now let $0 \neq \bar{y} \in R/A$ and suppose that $J\bar{y} = 0$ for some ideal $J \in T(\overset{0}{\tau}_{R/P})$, i.e., $J\, Ry \subset A$ and $J(A + Ry) \subset A$. Since σ is symmetric and $\sigma(R/P) = 0$ it is easily seen that $A + Ry \in T(\overset{0}{\tau}_{R/P})$. Now, $\overset{0}{\tau}_{R/P}$ being symmetric, we get that $J(A + Ry) \in T(\overset{0}{\tau}_{R/P})$ and $A \in T(\overset{0}{\tau}_{R/P})$. The latter yields $P = [A : R] \in T(\overset{0}{\tau}_{R/P})$, contradiction. Thus $\overset{0}{\tau}_{R/P} \leqslant \tau_{R/A}$.

COROLLARY 1. From Proposition 20 it follows that for a restricted functor σ, the elements A of $C'(\sigma)$ are such that $[A : R] = P \in C(\sigma)$. Therefore $\overset{0}{\tau}_{R/A} = \overset{0}{\tau}_{R/P}$.

COROLLARY 2. Let σ be a restricted kernel functor such that $\tau_{R/A}$ is symmetric for every $A \in C'(\sigma)$, then the fact that the quasi-prime kernel functors $\overset{0}{\tau}_{R/P}$ coincide for all $P \in C(\sigma)$ implies that σ is a prime kernel functor.

PROPOSITION 26. Let R be a left Noetherian prime ring, let P be a prime ideal of R. Suppose that σ_{R-P} is a T-functor such that the elements of G(P) are units in $Q_{R-P}(R)$, then :

1. $\sigma_P = \sigma_{R-P}$
2. R satisfies the left Ore condition with respect to G(P).
3. σ_{R-P} is a restricted quasi-prime.
4. σ_{R-P} is a prime kernel functor if and only if the induced $\tau_{R/A}$ are symmetric for all $A \in C'(\sigma)$.
5. The elements of $C'(\sigma_{R-P})$ are prime left ideals, they are the left ideals maximal with the property of being disjoint from G(P).

PROOF. 1. If $g \in G(P)$ then Rg is killed under extension to $Q_{R-P}(R)$ because g is a unit in $Q_{R-P}(R)$. Property (T) for σ_{R-P} then implies that $Rg \in T(\sigma_{R-P})$ and $T(\sigma_P) = T(\sigma_{R-P})$ follows because every $A \in T(\sigma_P)$ contains some $g \in G(P)$.

2. The left Ore condition for R with respect to G(P) is equivalent to the fact that left ideals Rg, $g \in G(P)$, are in $T(\sigma_P)$. The latter is a consequence of 1. above.

3. Proposition 17 may be applied; we get $P = \cap \{A, A \in C'(\sigma_P)\}$ and thus $\sigma_{R-P} = \sigma_P$ is restricted.

4. If σ_{R-P} is prime then all $\tau_{R/A}$, $A \in C'(\sigma)$, coincide and coincide with σ_{R-P} because $\sigma_{R-P} = \inf\{\tau_{R/A}, A \in C'(\sigma_{R-P})\}$. Conversely, let $\tau_{R/A}$ be symmetric for every $A \in C'(\sigma_{R-P})$. Since σ_{R-P} is restricted, Proposition 25 Corollary 2 apllies, yielding directly that σ_{R-P} is a prime kernel functor.

5. This is Proposition 20 Corollary 2.

The foregoing proposition also holds in case R is not prime; the proof then uses reduction techniques thus disposing of "torsion problems".

We mention the following :

1. If for some $A \in C'(\sigma_{R-P})$ it happens that $P \notin T(\tau^0_{R/A})$ then

$\sigma_{R-P} = \tau^0_{R/A}$. The converse is obviously also true.

PROOF. Since $\sigma_{R-P} \leq \tau^0_{R/A}$ and $\tau^0_{R/A}(R/P) \neq R/P$, Proposition 24 Corollary 1, entails that $\sigma_{R-P} = \tau^0_{R/A}$.

2. If σ_{R-P} is prime then $P \notin \dot{T}(\tau_{R/A})$ for every $A \in C'(\sigma)$, the converse is true when $\tau_{R/A}$ is symmetric for all $A \in C'(\sigma)$.

II. 3. Reductions.

In this section we have put together some properties of kernel functors and localization, with respect to changement of ground ring. From the localizer's point of view, it is most natural to transform R into $R/\sigma(R)$, therefore, although we have chosen a more general set up, many applications of the reduction-theory deal with the special case $R \to R/\sigma(R)$.

Let R_1 and R_2 be rings with unit and let $M(R_1)$ and $M(R_2)$ be the categories of R_1-modules and R_2-modules resp. To a $\sigma_i \in F(R_i)$ a filter T_i in R_i is associated, $i = 1,2$. A ring homomorphism $f : R_1 \to R_1$ is continuous if and only if $f^{-1}(T_2) \subset T_1$. A continuous and surjective ring homomorphism $f : R_1, T_1 \to R_2, T_2$ is said to be a final morphism if $f(T_1) \subset T_2$, i.e., if f is an open map of topological spaces. If f is a continuous homomorphism then an $M \in M(R_2)$ may be considered as an R_1-module via f and $\sigma_2(M) \subset \sigma_1(M)$ for all $M \in M(R_2)$. Suppose $f : R_1,T_1 \to R_2,T_2$ is onto and continuous and take $M_1 \in M(R_1)$, $M_2 \in M(R_2)$. A map $g : M_1 \to M_2$ is called a reduction (over f), or M_2 is said to be a reduction of M_1, if the following two properties hold :

1. g is R_1-linear. Hence g is continuous for the σ_i-topologies in M_i, $i = 1,2$.

2. $\sigma_2(M_2) \subset g(\sigma_1(M_1))$.

In case M_i is an R_i-ring, $i = 1,2$, given by ring homomorphisms

$j_i : R_i \to M_i$, then a <u>ring-reduction</u> $g : M_1 \to M_2$ is a reduction such that $g j_1 = j_2 f$.

<u>Examples</u>.

1. If $M = M_1 = M_2$, then the identity g of M is a reduction over f if and only if $\sigma_2(M) \subset \sigma_1(M)$. Hence, 1_M is a reduction over f for all $M \in M(R_2)$ if and only if $\sigma_2 \leq \sigma_1$ on $M(R_2)$.

2. Let $R_1 = R_2 = R$, $\sigma_1 = \sigma_2 = \sigma$ and $f = 1_R$. Let $M_1, M_2 \in M(R)$. An R-linear map $g : M_1 \to M_2$ is a reduction if and only if $\sigma(M_2) = g\,\sigma(M_1)$. If g is onto and Ker $g \subset \sigma(M_1)$ then we may deduce the following criterion : σ is idempotent if and only if for any $M \in M(R)$, all surjective R-linear maps g with Ker $g \subset \sigma(M)$ are reductions over 1_R.

Surjective reductions are called <u>epireductions</u>. A reduction of a σ_1-torsion free module is σ_2-torsion free. Obviously, composition of reductions yields a reduction over the composition of the underlying ring homomorphisms.

<u>PROPOSITION</u> 27. Let $f : R_1, T_1 \to R_2, T_2$ be a surjective continuous ring homomorphism. Let σ_1 be idempotent, then $f = f_0 \circ f_t$ where f_t is final and Ker $f_t \subset \sigma_1(R_1)$ while $\sigma_1(\text{Ker } f_0) = 0$.

<u>PROOF</u>. Since $\sigma_1(\text{Ker } f)$ is an ideal we may consider the following diagram of ring homomorphisms :

$$\begin{array}{ccccccccc}
0 & \longrightarrow & \text{Ker } f & \longrightarrow & R_1 & \xrightarrow{\;f\;} & R_2 & \to & 0 \\
 & & \Big\downarrow{\scriptstyle f_t} & & \Big\downarrow{\scriptstyle f_t} & & \Big\downarrow{\scriptstyle \cong} & & \\
0 & \to & \text{Ker } f/\sigma_1(\text{Ker } f) & \to & R_1/\sigma_1(\text{Ker } f) & \xrightarrow[f_0]{} & R_1/\text{Ker } f & \to & 0
\end{array}$$

Thus $\sigma_1(\text{Ker } f_0) = 0$ and since the filter in $R_1/\sigma_1(\text{Ker } f)$ is Im T_1 it follows that f_t is open and continuous while f_0 is continuous.

Proposition 28 will complete the proof, since there it is shown that f_t actually is a reduction.

A continuous surjective ring homomorphism $f : R_1, T_1 \to R_2, T_2$ such that $\mathrm{Ker}\, f \subset \sigma_1(R_1)$ is called a <u>torsion morphism</u>. A reduction $g : M_1 \to M_2$ with $\mathrm{Ker}\, g \subset \sigma_1(M_1)$ is a <u>torsion reduction</u>. If f is final, reductions over f are said to be <u>final reductions</u>.

All kernel functors are idempotent unless otherwise stated.

<u>PROPOSITION</u> 28. A final torsion morphism $f : R_1, T_1 \to R_2, T_2$ is a final torsion reduction of R_1-modules over f.

<u>PROOF</u>. Consider the exact sequence :

$$0 \to \mathrm{Ker}\, f \to f^{-1}(\sigma_2(R_2)) \to \sigma_2(R_2) \to 0.$$

Since $\mathrm{Ker}\, f \subset \sigma_1(R_1)$ while $\sigma_2(R_2)$ is also σ_1-torsion it follows that $f^{-1}(\sigma_2(R_2)) \subset \sigma_1(R_1)$ entailing that f is a reduction.

It is easily checked that if $f : R_1, T_1 \to R_2, T_2$ is final and if σ_1 is Noetherian, then σ_2 is also a Noetherian kernel functor (as defined in [12]).

<u>PROPOSITION</u> 29. Let $\sigma_1, \sigma_2 \in F(R)$ and suppose that σ_1 is idempotent. If there exists a final homomorphism $f : R_1, T_1 \to R_2, T_2$, then σ_2 is idempotent.

<u>PROOF</u>. Let A', B' be left ideals of R_2 such that $A' \in T_2$, $B' \subset A'$ and suppose that A'/B' is σ_2-torsion. Put $B = f^{-1}(B')$, $A = f^{-1}(A')$. Then $A/B \cong A'/B'$ and $A \in T_1$. Since A/B is σ_2-torsion it is also σ_1-torsion and thus the fact that σ_1 is idempotent implies $B \in T_1$ and $B' = f(B) \in T_2$ follows, proving that σ_2 is idempotent.

PROPOSITION 30. A final morphism $f : R_1, T_1 \to R_2, T_2$ gives rise to an embedding of $M(R_2)$ into $M(R_1)$ so that the restriction of σ_1 to $M(R_2)$ is equal to σ_2.

The proof is easy. Before investigating whether the restriction of the localization functor Q_1 to the embedded $M(R_2)$ equals Q_2, we include some reduction-properties of σ-projective modules.

PROPOSITION 31. Let $g : M_1 \to M_2$ be a torsion epireduction. Then, if M_1 is σ_1-projective, M_2 is σ_1-projective too.

PROOF. Given an exact sequence $M \to M'' \to 0$ of σ_1-torsion free R_2-modules and an R_2-linear map $h : M_2 \to M''$. Since M_1 is σ_1-projective we obtain a commutative diagram :

$$\begin{array}{ccccccccc}
0 & \longrightarrow & M_1' & \longrightarrow & M_1 & \longrightarrow & M_1/M_1' & \longrightarrow & 0 \\
 & & \Big\downarrow \psi & & \Big\downarrow g & & & & \\
 & & & & M_2 & & & & \\
 & & & & \Big\downarrow & & & & \\
 & & M & \longrightarrow & M'' & \longrightarrow & 0 & &
\end{array}$$

where M_1/M_1' is σ_1-torsion. Put $M_2' = g(M_1')$. Since M_1/M_1' maps onto M_2/M_2' the latter R_2-module is σ_1-torsion. Now $\text{Ker } g \subset \sigma_1(M_1)$, thus $M_1' \cap \text{Ker } g \subset \sigma_1(M_1')$, and thus $\psi(M_1' \cap \text{Ker } g)$ is σ_1-torsion in the σ_1-torsion free M. Therefore $M_1' \cap \text{Ker } g \subset \text{Ker } \psi$, or ψ factorizes through $g(M_1')$ and we obtain the following commutative diagram with exact rows :

$$\begin{array}{ccccccccc}
0 & \longrightarrow & M_1' & \longrightarrow & M_1 & \longrightarrow & M_1/M_1' & \longrightarrow & 0 \\
 & & \Big\downarrow & & \Big\downarrow g & & \Big\downarrow & & \\
0 & \longrightarrow & M_2' & \longrightarrow & M_2 & \longrightarrow & M_2/M_2' & \longrightarrow & 0 \\
 & & \Big\downarrow & & h \Big\downarrow & & \Big\downarrow & & \\
 & & M & \longrightarrow & M'' & \longrightarrow & 0 & &
\end{array}$$

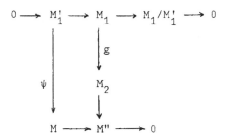

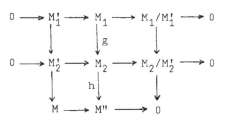

Since M_2/M_2' is σ_1-torsion, we may conclude that M_2 is σ_1-projective.

<u>COROLLARY</u>. If in the situation described above, g is final then σ_1-projectivity of M_1 yields σ_2-projectivity of M_2.

<u>PROPOSITION</u> 32. Let σ_1 be such that every left ideal in T_1 is σ_1-projective. Let $g : M_1 \to M_2$ be a final epireduction and let M_1 be faithfully σ_1-injective, then M_2 is faithfully σ_2-injective.

<u>PROOF</u>. Put $A = f^{-1}(A')$ for some $A' \in T_2$, hence $A \in T_1$. We have to show that any R_2-linear map $h : A' \to M_2$ extends to an R_2-linear $R_2 \to M_2$. By σ_1-projectivity of A we have a commutative diagram :

$$\begin{array}{ccccc} 0 \longrightarrow & B & \longrightarrow & A & \\ & \Big\downarrow & & \Big\downarrow f & \\ & \psi & & A' & \\ & \Big\downarrow & & \Big\downarrow & \\ & M_1 & \xrightarrow[g]{} & M_2 & \longrightarrow 0 \end{array}$$

where A/B is σ_1-torsion.
Because M_1 is faithfully σ_1-injective, ψ may be extended to an R_1-linear $\psi : R_1 \to M_1$ and thus $\psi(b) = b\,m$ for all $b \in B$ and some fixed $m \in M_1$. For any $a \in A$, $(h \circ f)(a) = g\psi(a) = g(am) = ag(m)$. Then, by definition of the R_1-module structure in M_2, this yields $h(a') = a'g(m)$, $a' \in A'$. Hence $h : A' \to M_2$ extends to an R_2-linear map $R_2 \to M_2$ which is defined by multiplication on the right by $g(m)$. Finally, M_2 is σ_2-torsion free, this proves that M_2 is faithfully σ_2-injective. Since a final torsion morphism is a torsion reduction, see Proposition 28, one could use Proposition 31 Corollary, to deduce conditions for the descent of property (T) in case R_1 is a left Noetherian ring. However, the following proposition will yield a similar result under milder hypotheses.

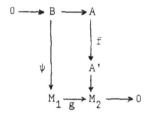

PROPOSITION 33. Let $f : R_1 \to R_2$ be a surjective ring homomorphism and let σ_1 be a T-functor on $M(R_1)$ such that R_1 and R_2 are both σ_1-torsion free. Equivalently :

1. The R_1-module structure of $Q_{\sigma_1}(R_2)$ defines an R_2-module structure for $Q_{\sigma_1}(R_2)$ via f.
2. Define σ_2 on $M(R)$ by putting $T(\sigma_2) = \{f(A), A \in T(\sigma_1)$, then $Q_{\sigma_1}(R_2) = Q_{\sigma_2}(R_2)$.
3. The unique extension $\tilde{f} : Q_{\sigma_1}(R_1) \to Q_{\sigma_1}(R_2)$ of f is a ring homomorphism.

PROOF. The surjectivity of f implies that σ_2 is indeed an idempotent functor, and it is clear that σ_1 and σ_2 coincide on R_2-modules. Suppose 1. The R_1-linear $i' : R_2 \to Q_{\sigma_1}(R_2)$ is R_2-linear and $Q_{\sigma_1}(R_2)/i'(R_2)$ is σ_1- and σ_2-torsion. Thus i' extends to a unique R_1-linear map

$$\psi : Q_{\sigma_2}(R_2) \to Q_{\sigma_1}(R_2),$$

while $i : R_2 \to Q_{\sigma_2}(R_2)$ extends to a unique R_2-linear map

$$\phi : Q_{\sigma_1}(R_2) \to Q_{\sigma_2}(R_2)$$

which is also R_1-linear by definition of the R_1-module structure of $Q_{\sigma_1}(R_2)$. Since $Q_{\sigma_1}(R_2)$ is easily seen to be σ_2-injective in $M(R_2)$ it follows that $Q_{\sigma_1}(R_2) \cong Q_{\sigma_2}(R_2)$. To prove that 2. implies 3., consider the commutative diagram of R_1-linear morphisms :

$$\begin{array}{ccc} R_1 & \xrightarrow{f} & R_2 \\ \downarrow & & \downarrow \\ Q_{\sigma_1}(R_1) & \xrightarrow{\tilde{f}} & Q_{\sigma_1}(R_2) \cong Q_{\sigma_2}(R_2). \end{array}$$

Give $g_\eta : R_1 \to Q_{\sigma_1}(R_1)$, by $g_\eta(r) = r\eta$, then g_η extends to a unique

$f_\eta : Q_{\sigma_1}(R_1) \to Q_{\sigma_1}(R_1)$. Now, $\tilde{f}f_\eta(r) = \tilde{f}(r\,\eta) = r\,\tilde{f}(\eta) = f(r)\tilde{f}(\eta)$, and on the other hand $f_{\tilde{f}(\eta)}\tilde{f}(r) = \tilde{f}(r)\tilde{f}(\eta) = f(r)\tilde{f}(\eta)$. Hence, $\tilde{f}f_\eta$ and $f_{\tilde{f}(\eta)}\tilde{f}$ coincide on R_1, therefore these maps coincide on $Q_{\sigma_1}(R_1)$.
If $\xi,\eta \in Q_{\sigma_1}(R_1)$, then $\tilde{f}(\xi\eta) = \tilde{f}f_\eta(\xi) = f_{\tilde{f}(\eta)}\tilde{f}(\xi) = \tilde{f}(\xi)\tilde{f}(\eta)$. Finally, if $\tilde{f}$ is a ring homomorphism then $\tilde{f}(r\eta) = \tilde{f}(r)\tilde{f}(\eta) = f(r)\tilde{f}(\eta) = r\tilde{f}(\eta)$, the latter equality, follows by R_1-linearity of $\tilde{f}$. Since property (T) implies that $\tilde{f}$ is surjective, the implication 3 ⇒ 1 follows.

COROLLARY 1. Any of the conditions in Proposition 33 implies that Ker f extends to an ideal $Q_{\sigma_1}(\text{Ker } f)$ of $Q_{\sigma_1}(R_1)$ and this again is equivalent to statement 3. Note that the implications 1 ⇒ 2 ⇒ 3 hold even without the assumption that σ_1 has property (T).

COROLLARY 2. In case Ker f $= \tau(R_1)$ for some (symmetric) T-functor $\tau \geq \sigma_1$, then Theorem 10 yields that Ker f is a σ_1-ideal and therefore the foregoing proposition applies. Hence we obtain a ring homomorphism

$$\tilde{f} : Q_{\sigma_1}(R_1) \to Q_{\sigma_1}(R_1/\tau(R_1)),$$

which is onto. Before deriving Corollary 3 we turn to the following, more general situation.

Reduction of property (T).

Let R be an arbitrary ring and let τ be a T-functor on M(R). Consider $j_\tau : R \to R/\tau(R)$. If $R/\tau(R)$ is equiped with the filter $j_\tau T(\tau)$ then j_τ is a final torsion morphism. Hence, if A is a left ideal of R then $A \to j_\tau(A)$ is a final torsion epireduction. Since τ has property (T) it follows that τ is Noetherian and moreover, every $B \in T(\tau)$ contains an $A \in T(\tau)$ which is τ-projective.
Consequently, $j_\tau(\tau)$ is Noetherian and the corollary to Proposition 31 yields that every $j_\tau(\tau)$-open left ideal contains a $j_\tau(\tau)$-projective left ideal which also is in $T(j_\tau(\tau))$. Hence, $j_\tau(\tau)$ is a T-functor.

COROLLARY 3. We may restate Proposition 33 under milder assumptions. Let $f : R_1 \to R_2$ be a surjective ring homomorphism and let σ_1 be a T-functor on $M(R_1)$, then the following assertions are equivalent :

1. $Q_{\sigma_1}(R_2)$ is an R_2-module via f
2. If $\sigma_2 = f(\sigma_1)$ then $Q_{\sigma_1}(R_2) = Q_{\sigma_2}(R_2)$
3. The extension $\tilde{f} : Q_{\sigma_1}(R_1) \to Q_{\sigma_1}(R_2)$ is a ring homomorphism.

PROOF. We reduce this to the proof of Proposition 33 as follows. Consider the following commutative diagram of ring homomorphisms

$$\begin{array}{ccc} R_1 & \xrightarrow{\ f\ } & R_2 \\ \Big\downarrow{\scriptstyle \pi_1} & & \Big\downarrow{\scriptstyle \pi_2} \\ R_1/\sigma_1(R_1) & \xrightarrow[f^*]{} & R_2/\sigma_1(R_2) \end{array}$$

Equiping $\pi_1(R_1)$ and $\pi_2(R_2)$ with the residue filters of $T(\sigma_1)$ and $T(\sigma_2)$, we obtain a diagram of final morphisms with vertical arrows representing torsion morphisms. The facts listed below reduce statement 1,2,3 to analogous statements in the level of f^* and then Proposition 33 applies.

1. The R_1-module structure of $Q_{\sigma_1}(R_2)$ is defined via π_1.
2. $Q_{\sigma_1}(R_2) = Q_{\sigma_1}(R_2/\sigma_1(R_2)) = Q_{\sigma_1}(R_2/\sigma_2(R_2))$
3. Commutativity of the diagram yields $\pi_2\, f\, T(\sigma_1) = \pi_2\ T(\sigma_2) = f^*\pi_1 T(\sigma_1)$
4. The extension $\tilde{f} : Q_{\sigma_1}(R_1) \to Q_{\sigma_1}(R_2)$ of f to $Q_{\sigma_1}(R_1)$ is actually the extension of f^*, by definition.

Another consequence of the reduction of property (T) is :

PROPOSITION 34. Let $f : R_1, T_1 \to R_2, T_2$ be a final torsion morphism. Let σ_1 be a T-functor, let Ker f be a σ_1-ideal of R_1. Then $Q_{\sigma_1}(R_2) = Q_{\sigma_2}(R_2)$. For every $M \in M(R_2)$ we have that $Q_{\sigma_1}(M) \cong Q_{\sigma_2}(M)$, i.e., the localization functors Q_{σ_1} and Q_{σ_2} coincide on R_2-modules.

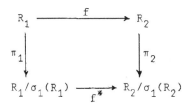

PROOF. By property (T) for σ_1 we obtain that $Q_{\sigma_1}(M) \cong Q_{\sigma_1}(R_1) \otimes_{R_1} M$ and descent of property (T) under f yields that $Q_{\sigma_2}(M) = Q_{\sigma_2}(R_2) \otimes_{R_2} M$. From Proposition 33 we retain that

$$Q_{\sigma_1}(R_2) = Q_{\sigma_2}(R_2) \text{ and thus } Q_{\sigma_1}(R_1) = Q_{\sigma_2}(R_2)$$

(since $Q_{\sigma_1}(R_2) = Q_{\sigma_1}(R_1)$ because f is torsion). Finally, the fact that Ker f is a σ_1-ideal entails that both M and $Q_{\sigma_1}(R_1)$ are R_2-modules and R_1-modules via f, thus the tensor products are isomorphic.

If $g : M_1 \to M_2$ is a final torsion reduction then $g^{-1}(\sigma_2(M_2)) = \sigma_1(M_1)$ and g induces a final torsion reduction $g\sigma_1 : \sigma_1(M_1) \to \sigma_2(M_2)$. If g is a final torsion epireduction then $M_1/\sigma_1(M_1) \cong M_2/\sigma_2(M_2)$. This is an easy consequence of the fact that we have an exact sequence

$$0 \to \text{Ker } g \to g^{-1}(\sigma_2(M_2)) \to gg^{-1}(\sigma_2(M_2) \to 0$$

because $g^{-1}(\sigma_2(M_2)) \subset \sigma_1(M_1)$ and $g(\sigma_1(M_1)) \subset \sigma_1(M_2)$ imply that $g\sigma_1 = g|\sigma_1(M_1)$ is a final torsion reduction and if g is onto then the isomorphism $M_1/\sigma_1(M_1) \cong M_2/\sigma_2(M_2)$ follows. To a final reduction $g : M_1 \to M_2$ there corresponds a reduction $Q_{\sigma_1}g : Q_{\sigma_1}(M_1) \to Q_{\sigma_2}(M_2)$. This map is not necessarily a reduction of $Q_{\sigma_1}(R_1)$-modules. If σ_1 is a T-functor while Ker f is a σ_1-ideal then $Q_{\sigma_1}g$ is a reduction of $Q_{\sigma_1}(R_1)$-modules because in this case the R_1-linear map $Q_{\sigma_1}(R_1) \to Q_{\sigma_2}(R_2)$ is onto and a ring homomorphism.

LEMMA 35. Let $f : R_1, T_1 \to R_2, T_2$ be a final morphism. Then σ_1 is symmetric if and only if σ_2 is symmetric, and if so then :

1. Taking inverse images under f yields an injection $C'(\sigma_2) \to C'(\sigma_1)$.
2. Taking inverse images under f yields an injection $C(\sigma_2) \to C(\sigma_1)$.

PROOF. The first statement is trivial. To prove 1, let $A' \in C(\sigma_2)$.

Then $A = f^{-1}(A') \notin T_1$ since f is open and also, $A \in C'(\sigma_1)$ because if $B \supset A$ with $B \notin T_1$ then $B \supset \mathrm{Ker}\, f$ yields $B = f^{-1}f(B)$ where $f(B) \supset A'$. Hence $f(B) = A'$ follows and then B has to coincide with A. The proof of 2, is similar.

PROPOSITION 36. Let $f : R_1, T_1 \to R_2, T_2$ be a final torsion morphism and let σ_1 be a T-functor such that Ker f extends to an ideal $Q_{\sigma_1}(\mathrm{Ker}\, f)$ of $Q_{\sigma_1}(R_1)$, then σ_1 is a prime kernel functor if and only if σ_2 is.

PROOF. Note that every $A \in C'(\sigma_1)$ is saturated because $A \supset \sigma_1(R_1) \supset \mathrm{Ker}\, f$. Thus $A \to f(A)$ sets up a one-to-one correspondence between the sets $C'(\sigma_1)$ and $C'(\sigma_2)$. The corollary to Proposition 8 yields that σ_1 is prime if and only if all R_1-modules $Q_{\sigma_1}(R_1/A)$ with $A \in C'(\sigma_1)$ are isomorphic to one another. Let $A,B \in C'(\sigma_1)$ and write $A' = f(A)$, $B' = f(B)$, then $A',B' \in C'(\sigma_2)$. We have $Q_{\sigma_1}(R_2/A') \cong Q_{\sigma_1}(R_1/A) \cong Q_{\sigma_1}(R_1/B) \cong Q_{\sigma_1}(R_2/B')$. Proposition 34 implies that $Q_{\sigma_2}(R_2/A') \cong Q_{\sigma_2}(R_2/B')$ and since this holds for arbitrary $B' \in C'(\sigma_2)$ we derive from this that σ_1 is prime if and only if σ_2 is prime.

Let $f : R_1, T_1 \to R_2, T_2$, be a continuous and surjective ring homomorphism. One easily checks the following elementary properties which are of pure ring theoretic nature. If A and B are left ideals of R_1 such that $A \supset \mathrm{Ker}\, f$ then,

$$f[A : B] = [f(A) : f(B)] \text{ and } [A : B] = f^{-1}[f(A) : f(B)].$$

If A and B are left ideals of R_2 then, $[f^{-1}(A) : f^{-1}(B)] = f^{-1}[A : B]$. If f is final and if $A \in C'(\sigma_1)$ is such that $A \supset \mathrm{Ker}\, f$ then, $f(A) \in C'(\sigma_2)$. If $P \in C(\sigma_1)$, $P \supset \mathrm{Ker}\, f$ then $f(P) \in C(\sigma_2)$.

PROPOSITION 37. Let $f : R_1, T_1 \to R_2, T_2$ be a final morphism.

1. If σ_1 is restricted then σ_2 is restricted.
2. If f is a final torsion reduction then σ_1 is restricted if and only if σ_2 is restricted.

PROOF. 1. Let $A \in C'(\sigma_2)$ then $f^{-1}(A) \in C'(\sigma_1)$ and $[f^{-1}(A) : R_1] \in C(\sigma_1)$. Thus $f[f^{-1}(A) : R_1] = [A : R_2]$ and $[A : R_2] \in C(\sigma_2)$ by the foregoing remarks.

2. Since $\mathrm{Ker}\, f \subset \sigma_1(R_1)$, every $A \in C'(\sigma_1)$ contains Ker f and hence if $A \in C'(\sigma_1)$ then $f(A) \in C'(\sigma_2)$. Moreover $[f(A) : R_2] \in C(\sigma_2)$ yields that $f^{-1}[f(A) : R_2] \in C(\sigma_1)$ and thus $[f^{-1}f(A) : R_1] \in C(\sigma_1)$. Thus if σ_2 is restricted then so is σ_1.

Conclusion. If one considers $R \to R/\sigma(R)$ for an arbitrary idempotent $\sigma \in F(R)$, then most properties of σ and Q_σ on $M(R)$ are equivalent to the analogous properties for the restriction $\bar{\sigma}$ of σ to the embedded category of $R/\sigma(R)$-modules.

Special references for Section II.

A.W. GOLDIE [11]; A.G. HEINICKE [13]; J. LAMBEK, G. MICHLER [21]; D.C. MURDOCH, F. VAN OYSTAEYEN [25], [26], [27]; S.K. SIM [30], [31]; F. VAN OYSTAEYEN [38].

III. SHEAVES

III. 1. Spec and the Zariski Topology.

R is a left Noetherian ring with unit.
Put $X = \mathrm{Spec}\, R = \{\text{proper prime ideals of } R\}$. To any ideal A of R the set $V(A) = \{P \in X,\ P \supset A\}$ is associated, this set obviously depends only on the radical rad A of A. The following is clear.

LEMMA 38.

1. Let A, B be ideals of R and let $A \subset B$, then $V(A) \supset V(B)$.
2. For a set $\{A_\alpha,\ \alpha \in I\}$ of ideals of R we have that $V(\sum_\alpha A_\alpha) = \bigcap_\alpha V(A_\alpha)$.
3. For ideals A, B of R, $V(A \cap B) = V(AB) = V(A) \cup V(B)$.

So we may take the sets $X_A = X - V(A)$ to be the open sets for a topology on X, called the Zariski topology. A point P of X is closed if and only if P is a maximal ideal of R. A subset $S \subset X$ is said to be irreducible if S is not the union of closed sets which are different from S. A generic point for an irreducible set S is a $P \in X$ such that $V(P) = S$.

PROPOSITION 39. If $P \in X$, then $V(P)$ is irreducible. Conversely, every irreducible closed subset $S \subset X$ is equal to $V(P)$ for some $P \in X$ and P is the unique generic point for S.

PROOF. If $V(P) = W_1 \cup W_2$, where W_1 and W_2 are closed sets, then P is one of these sets, $P \in W_1$. Hence $W_1 = V(P)$. Conversely, let $V(A)$ be irreducible. Since $V(A) = V(\mathrm{rad}\, A)$ we assume $A = \mathrm{rad}\, A$. If A is not prime then there exist ideals B' and C' which are not contained in A such that $B'C' \subset A$. Put $B = A + B'$, $C = A + C'$. Then $A = B \cap C$; indeed, if $x \in B \cap C$ then $x = a_1 + b = a_2 + c$ with $a_1, a_2 \in A$ and $b \in B$, $c \in C$. Hence, $x R x \subset b R c + A \subset A$ and this yields $x \in A$ since A is radical.

Therefore, $V(A) = V(B) \cup V(C)$ and for any $x \notin A$ there exists a $P \in V(A)$ such that P does not contain the ideal (x). Hence, picking an $x \in B - A$, we get that $V(A) - V(B) \neq \phi$ entailing a contradiction with the irreducibility of $V(A)$. If P is another generic point for $V(A)$ then $P \supset A$ but since $A \in V(P)$ also $A \supset P$, whence $A = P$.

As in the case of a commutative ring it is easily verified that a set $\{A_\alpha, \alpha \in I\}$ of ideals of R gives rise to a covering of X by means of open sets $X_\alpha = X - V(A_\alpha)$ if and only if $1 \in \sum_\alpha A_\alpha$. It follows that X is compact but not necessarily Hausdorff; for any ideal A of R, the open set X_A is compact.

Let A,B be ideals of R, then $A \subset \text{rad } B$ is equivalent to $A^n \subset B$ for some positive integer n. To an ideal A of R we associate a filter,

$T(A) = \{$left ideals L of R containing an ideal B such that $\text{rad } B \supset A\}$

$= \{$left ideals L of R such that $L \supset A^n$ for some positive integer $n\}$.

Obviously, $T(A)$ defines a symmetric kernel functor σ_A,

$\sigma_A(M) = \{m \in M,\ Lm = 0$ for some $L \in T(A)\}$ for every R-module M.

Note that the set $C(\sigma_A)$ is a subset of X_A and it may be looked upon as being the set of "tops" of X_A, i.e. the maximal elements of X_A. Using the localization technique expounded in section II, we associate to any non-empty open set X_A, i.e. A not contained in rad (0), the ring of quotients $Q_A(R)$ with respect to σ_A. This is well defined because both X_A and $T(A)$ depend only on the radical of A. The canonical maps $R \to Q_A(R)$ are denoted by i_A, though in general they are not injective.

PROPOSITION 40. Assigning $Q_A(R)$ to X_A for every ideal A not contained in rad (0), defines a presheaf of noncommutative rings on Spec R.

PROOF. If $X_B \subset X_A$ then $\text{rad } B \subset \text{rad } A$ and $X_B \neq \phi$ if and only if $B \not\subset \text{rad } (0)$. Pick an L in $T(A)$, then $L \supset A^n$ and since $A \supset B^m$ we obtain $L \supset B^{mn}$ and $L \in T(B)$. This proves the inclusion $T(A) \subset T(B)$. Thus

$\sigma_A \leqslant \sigma_B$ and we get the canonical projection π,

$$\pi(A,B) : i_A(R) = R/\sigma_A(R) \to R/\sigma_B(R) = i_B(R).$$

Now, consider the following diagram (all maps in the diagram are ring homomorphisms by Theorem 10)

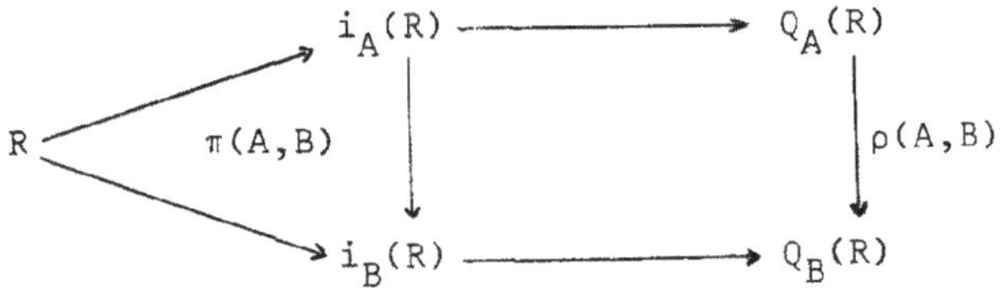

Since $Q_A(R)/i_A(R)$ is σ_A-torsion, a fortiori σ_B-torsion, the fact that $Q_B(R)$ is faithfully σ_B-injective implies that $\pi(A,B)$ extends uniquely to $\rho(A,B)$, and $\operatorname{Ker} \pi = \operatorname{Ker} \rho(A,B) \cap i_A(R)$. The uniqueness property also yields that $\rho(A,A)$ is the identity on $Q_A(R)$. If $\phi \neq X_C \subset X_B \subset X_A$, then the following diagram of ring homomorphisms results :

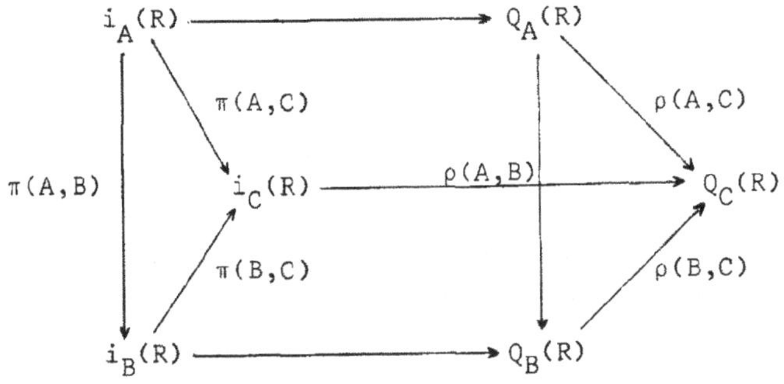

Since $\pi(B,C)\pi(A,B) = \pi(A,C)$, it follows that $\rho(A,C)$ and $\rho(B,C)\rho(A,B)$ are extensions of $\pi(A,C)$ to $Q_A(R)$ and as such they must coincide. Consequently, the diagram is commutative.

Remark. Let $\{A_i, i \in I\}$ be a finite set of ideals of R, then it is readily verified that the radical of $\sum_i A_i$ is equal to the radical of $\sum_i \operatorname{rad} A_i$.

<u>THEOREM</u> 41. The presheaf $(X_A, Q_A(R))$ denoted $\tilde{R}$ is a monopresheaf on X, i.e., if $X_A = \cup X_i$ is a covering of X_A by open sets $X_i = X - V(A_i)$ then $g = 0$ is the unique element of $Q_A(R)$ such that $\rho(A,A_i)g = 0$ for all i.

<u>PROOF</u>. Since X_A is compact, we may suppose that we are given a finite covering of X_A. Writing σ_j for the kernel functor corresponding to A_j, we obtain the following commutative diagram of ring homomorphisms

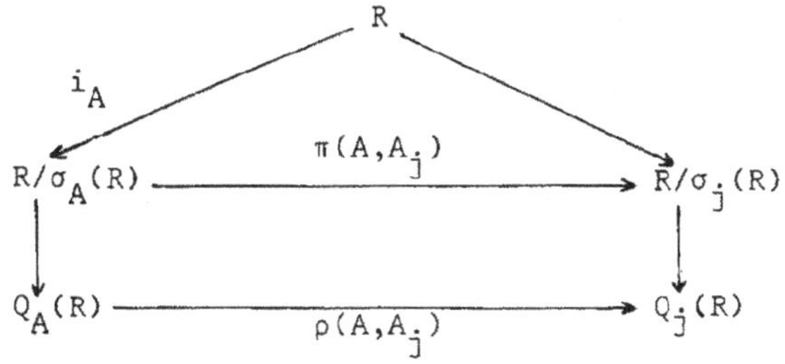

By definition of $Q_A(R)$ there exists an ideal $B \in T(A)$ such that $Bg \subset i_A(R)$. So if $g \in \mathrm{Ker}\,\rho(A,A_j)$ then $Bg \subset i_A(R) \cap \mathrm{Ker}\,\rho(A,A_j) = \mathrm{Ker}\,\pi(A,A_j)$, hence $Bg \subset \sigma_j(R)/\sigma_A(R)$ for all j. If we prove that $\bigcap_j \sigma_j(R)/\sigma_A(R) = 0$ then $Bg = 0$ and $g \in \sigma_A(Q_A(R)) = 0$ follows. Obviously, $\bigcap_j \sigma_j(R)$ contains $\sigma_A(R)$. Conversely, let $x \in \bigcap_j \sigma_j(R)$, then for each j there is an ideal $C_j \in T(A_j)$ such that $C_j\, x = 0$. By definition of σ_j we have that $\mathrm{rad}\, C_j \supset A_j$. Take $C = \sum_j C_j$. Then rad C is equal to $\mathrm{rad}(\sum_j \mathrm{rad}\, C_j)$ and contains $\mathrm{rad}(\sum_j A_j)$. But then the fact that $V(A) = \bigcap_j V(A_j)$ implies that any $P \in X$ containing $\sum_j \mathrm{rad}\, A_j$, hence containing $\mathrm{rad}\, A_j$ for every j, is in $V(A)$. Hence $\mathrm{rad}\, A \subset \mathrm{rad}(\sum_j A_j)$ and therefore C is in $T(A)$. Finally $C\, x = 0$ yields $x \in \sigma_A(R)$. Using classical sheafification methods we might construct a sheaf of sections associated with the presheaf. Probably this would lead to a theory similar to the theory of schemes over a commutative ring. However, in the present context we contend ourselves with the close generalization of affine schemes. Therefore, we make the further assumption that R is a prime ring, and prove that in this case the monopresheaf actually is a sheaf of rings.

THEOREM 42. Let R be a left Noetherian prime ring. The monopresheaf $\tilde{R}$ on Spec R is a sheaf, i.e., if we are given any covering of an open set X_A, $X_A = \bigcup_\alpha X_\alpha$, with $X_\alpha = X - V(A_\alpha)$, such that there exist elements $g_\alpha \in Q_\alpha(R)$ for which $\rho(A_\alpha, A_\alpha A_\beta)g_\alpha = \rho(A_\beta, A_\alpha A_\beta)g_\beta$, then there exists an element $g \in Q_A(R)$ such that $\rho(A, A_\alpha)g = g_\alpha$ for every α.

PROOF. Note first that is sufficient to prove the theorem for a finite covering of X_A. Indeed, if $X_A = \bigcup_i X_i$ is a finite covering of X_A for which the statement is true, then X_α is covered by the sets $X_\alpha \cap X_i = X - V(A_\alpha\ A_i)$. Let $g \in Q_A(R)$ map onto $g_i \in Q_i(R)$ and let h_α be the image of g in $Q_\alpha(R)$. Then g_α and h_α have the same image under every map $\rho(A_\alpha, A_\alpha A_i)$ and the foregoing theorem yields that $h_\alpha - g_\alpha = 0$, entailing the desired property for the arbitrary covering. So suppose that $X_A = \bigcup_i X_i$ is a finite covering for X_A. If the $g_i \in Q_i(R)$ have the prescribed property then we may derive from the presheaf axioms that : $\rho(A_i, \prod_i A_i)g_i = \rho(A_j, \prod_i A_i)g_j$ for all i and j, where $\prod_i A_i$ is a non-zero ideal of R because the product is finite. Note that the exact order of the factors in $\prod_i A_i$ is not important because everything is only depending on radicals. Denote $\prod_i A_i$ by B. We are going to prove that, if $g_1 \in Q_1(R)$ and $g_2 \in Q_2(R)$ are such that $\rho(A_1,B)g_1 = \rho(A_2,B)g_2$, then there is an element $g \in Q_C(R)$, where $C = A_1 + A_2$, such that $\rho(C,A_1)g = g_1$ and $\rho(C,A_2)g = g_2$. The theorem then follows easily if one repeates this process a finite number of times, because σ_A is exactly the kernel functor associated with $\sum_i A_i$. Elements of $Q_1(R)$ are defined to be equivalence classes $[L_1,f_1]$ of pairs (L_1,f_1) with $L_1 \in T(A_1)$ and $f_1 \in \mathrm{Hom}_R(L_1,R)$ the equivalence relation is given by : $(L_1,f_1) \sim (L_1',f_1')$ if and only if there is an $L_1'' \in T(A_1)$ such that f_1 and f_1' coincide on $L_1'' \subset L_1' \cap L_1$.

Put $g_1 = [L_1,f_1]_1 \in Q_1(R)$, $g_2 = [L_2,f_2]_2 \in Q_2(R)$. Because R is a prime ring, all the maps $\pi(A_i,B)$ reduce to the identity on R and thus it shows that every $\rho(A_i,B)$ is injective. These maps are then defined as follows :

$$\rho(A_1,B)[L_1,f_1]_1 = [L_1,f_1]_B$$

$$\rho(A_2,B)[L_2,f_2]_2 = [L_2,f_2]_B,$$

the right hand sides denoting the classes defined by the respective pairs but for the T(B)-equivalence relation. Since g_1 and g_2 map onto the same element of $Q_B(R)$ it follows that (L_1,f_1) is T(B)-equivalent to (L_2,f_2), i.e., $f_1|L' = f_2|L'$ for some $L' \subset L_1 \cap L_2$ and with $L' \in T(B)$, the latter may be taken to be an ideal because σ_B is symmetric. Pick an $x \in L_1 \cap L_2$. Then $L'x \subset L'$ yields $L'(f_1(x) - f_2(x)) = 0$ and thus $f_1(x) - f_2(x) \in \sigma_B(R) = 0$. Consequently f_1 and f_2 coincide on the whole of $L_1 \cap L_2$. Define $f : L_1 + L_2 \to R$ by $f(a_1 + a_2) = f_1(a_1) + f_2(a_2)$ with $a_1 \in L_1$, $a_2 \in L_2$. This is possible because $f_1|L_1 \cap L_2 = f_2|L_1 \cap L_2$. Take $g = [L_1 + L_2, f]_C$ with $C = A_1 + A_2$. Simple verification of the fact that g is $T(A_1)$-equivalent to $[L_1,f_1]_1$ and $T(A_2)$-equivalent to $[L_2,f_2]$ proves that g maps onto g_1,g_2 under $\rho(C,A_1)$, $\rho(C,A_2)$ resp.

DEFINITION. Let R be a left Noetherian prime ring. Spec R with the Zariski topology and the sheaf $\tilde{R}$, called the structure sheaf on Spec R, is said to be an affine scheme.

Let σ^* be the symmetric kernel functor defined by $C(\sigma^*) = \{(0)\}$, i.e., $\sigma^* = \sup\{\sigma_A$, A a nonzero ideal of R$\}$.

Let M be a σ^*-torsion free R-module. Then M is σ_A-torsion free for every σ_A, hence every nonzero submodule of M is faithful. Assigning $Q_A(M)$ to X_A defines a sheaf $\tilde{M}$ of R-modules on Spec R. The proof of the fact that $\tilde{R}$ is a sheaf may easilly be modified so as to apply to the module case.

III. 2. Affine Schemes.

Throughout this section, R is a left Noetherian prime ring. Let $X = \text{Spec } R$ be an affine scheme.
It is obvious that, for ideals A_1 and A_2 of R, we have that $T(A_1A_2) = \sup\{T(A_1), T(A_2)\}$. We agree to call $Q_{\sigma^*}(R)$ the function ring of Spec R and since $\sigma_A \leqslant \sigma^*$ for every ideal A of R we have injections $Q_A(R) \to Q_{\sigma^*}(R)$. It follows from these considerations that $Q_{\sigma^*}(R)$ may be considered as being the direct limit of the directed system $\{Q_A(R), \rho(A,B), 0 \neq B \subset A\}$. The following proposition determines the stalks of the structure sheaf.

PROPOSITION 43. Let $P \in X$, then :

1. $\sigma_{R-P} = \sup\{\sigma_A, P \in X_A\}$
2. $Q_{R-P}(R) = \varinjlim_{P \in X_A} Q_A(R)$.

PROOF. 1. If $P \in X_A$ then $P \not\supset A$, hence $A \supset (s)$ for some $s \in R-P$. The latter entails $\sigma_A \leqslant \sigma_{R-P}$. Conversely, if $B \in T(\sigma_{R-P})$ then $B \supset (s)$ for some s not in P. Hence, putting $A = (s)$ we have $P \in X_A$ and $B \in T(A)$.

2. This is an easy consequence of 1, since if $P \in X_A$ and $P \in X_B$ then $P \in X_{AB}$. It has been noted that $\sigma_{AB} = \sup\{\sigma_A,\sigma_B\}$. Furthermore, if the rings $Q_A(R)$ and $Q_B(R)$ are mapped injectively into some $Q_C(R)$ then also $Q_{AB}(R)$ maps injectively into $Q_C(R)$. The direct limit of the system

$$\{Q_A(R), \rho(A,B), 0 \neq B \subset A, P \in X_B\}$$

is obtained as the quotient ring for the symmetric functor $\sup\{\sigma_A, P \in X_A\} = \sigma_{R-P}$.
The ring $Q_{R-P}(R)$ is called the stalk of $\tilde{R}$ at P. The stalks of the

structure sheaf are mapped injectively into the function ring $Q_\sigma*(R)$. Similar to Proposition 43, we have that the stalks of the sheaf $\tilde{M}$ are given by the modules $Q_{R-P}(M)$. <u>Morphisms of affine schemes</u> are defined to be morphisms of the sheaf of rings, i.e., let R_1, R_2 be prime left Noetherian rings and let $X = \text{Spec } R_1$, $Y = \text{Spec } R_2$ be the corresponding affine schemes; a morphism from X to Y is then given by :

1. A Zariski continuous map $\tilde{\phi} : X \to Y$, such that
2. For every non-empty open set U in Y, write $\tilde{\phi}^{-1}(U) = V$ and let σ_U, σ_V be the associated kernel functors on $M(R_2)$ and $M(R_1)$ respectively, then there exist ring homomorphisms $\phi_U : Q_U(R_2) \to Q_V(R_1)$ which are compatible with sheaf restrictions.

A morphism of affine schemes is said to be an <u>isomorphism</u> if and only if $\tilde{\phi}$ is a homeomorphism of Zariski topological spaces and all induced maps ϕ_U are ring isomorphisms.

The functorial properties Spec enjoys in the commutative case cannot be obtained in full generality here. Restricting attention to special open sets and the kernel functors associated to them, we obtain a satisfactory local theory.

<u>Conventions and definitions</u> : An open set X_A in $X = \text{Spec } R$ is a <u>T-set</u> if σ_A is a T-functor; the stalk $Q_{R-P}(R)$ is a <u>T-stalk</u> if σ_{R-P} has property (T). If a T-functor σ_A(T-set X_A, T-stalk $Q_{R-P}(R)$) is such that R is σ_A-perfect (σ_A-perfect, σ_{R-P}-perfect) then it is called a <u>geometric functor</u> (set, stalk). In case R is a commutative Noetherian integral domain, all basic open sets $X_{(f)}$, $f \in R$, are T-sets and all stalks are T-stalks.

If R is not an hereditary ring however, not every open X_A is a T-set. For example if R is Noetherian, integrally closed but not a Dedekind domain, then, taking M to be a non-invertible maximal ideal in R we find that σ_M is not a T-functor (cf. [12], example 2, p. 45). It may sometimes be useful to have enough T-sets in Spec R. An affine scheme

Spec R is said to have a <u>T-basis</u> if there exists a basis for the Zariski topology, consisting of T-sets. Let M be a σ^*-torsion free R-module, and denote by $\tilde{M}$ the corresponding sheaf of R-modules on Spec R. A presheaf $\tilde{R} \otimes M$ may be defined by assigning $Q_A(R) \underset{R}{\otimes} M$ to X_A, we have :

<u>PROPOSITION</u> 44. If X = Spec R posseses a T-basis then $\tilde{M} \cong \Gamma(\tilde{R} \otimes M)$; $\Gamma(\tilde{R} \otimes M)$ being the sheaf of sections of the presheaf $\tilde{R} \otimes M$.

<u>PROOF</u>. Let X_A be a T-set. Then $Q_A(M) \cong Q_A(R) \underset{R}{\otimes} M$. A section $S \in \Gamma(X_A, \tilde{R} \otimes M)$ may thus be identified with a section in $\Gamma(X_A,\tilde{M})$. Since $\Gamma(X_A,\tilde{M}) \cong Q_A(M)$ and since both $\tilde{M}$ and $\tilde{R} \otimes M$ coincide on a basis for the topology in X it follows that $\tilde{M}$ is isomorphic to the sheafification of $\tilde{R} \otimes M$, hence $\tilde{M} \cong \Gamma(\tilde{R} \otimes M)$.

Note that, since the modules considered are to be σ^*-torsion free, it is impossible to deduce from the foregoing that σ_{R-P} is a T-functor for all $P \in X$. This will be proved later, see Proposition 49 corollary.

<u>PROPOSITION</u> 45. A geometric stalk is a "local" ring, i.e., it is a left Noetherian prime ring with a unique maximal ideal. The proof, using Proposition 11, is easy.

If σ^* is a T-functor then $Q_{\sigma^*}(R)$ has (0) for a maximal ideal and hence it is a simple ring, generalizing the function field of a variety over a commutative ring.

We return to a more general setting. Let $\tau \geq \sigma$ be symmetric kernel functors and suppose that σ is a T-functor. Let $T(\tau^e)$ be the set of extended left ideals A^e, $A \in T(\tau)$. Then, $T(\tau^e)$ is obviously the filter of an idempotent kernel functor τ^e on $M(Q_\sigma(R))$. With these conventions :

<u>LEMMA</u> 46. $Q_{\tau^e}(Q_\sigma(R)) \cong Q_\tau(Q_\sigma(R)) \cong Q_\tau(R)$.

PROOF. Consider $Q_\sigma(R)$-modules as R-modules via $R \hookrightarrow Q_\sigma(R)$. If $M \in M(Q_\sigma(R))$ then $x \in \tau^e(M)$ if and only if $A^e x = 0$ for some $A^e \in T(\tau^e)$, if and only if $A_\sigma x = 0$, $A_\sigma \in T(\tau)$, or equivalently $x \in \tau(M)$. Thus τ and τ^e coincide on $Q_\sigma(R)$-modules. The $Q_\sigma(R)$-module $Q_{\tau^e}(Q_\sigma(R))/Q_\sigma(R)$ is τ^e-torsion, hence τ-torsion as an R-module and an R-linear injective map ϕ results, $\phi : Q_{\tau^e}(Q_\sigma(R)) \hookrightarrow Q_\tau(Q_\sigma(R))$. Now, since $Q_{\tau^e}(Q_\sigma(R))$ is τ^e-injective, while $Q_\tau(Q_\sigma(R))$ is τ^e-torsion free and

$$Q_\tau(Q_\sigma(R))/Q_{\tau^e}(Q_\sigma(R))$$

being τ^e-torsion, a $Q_\sigma(R)$-module isomorphism : $Q_{\tau^e}(Q_\sigma(R)) \cong Q_\tau(Q_\sigma(R))$ follows. The fact that it is a ring isomorphism follows easily after verification of the fact that $Q_{\tau^e}(Q_\sigma(R))$ is faithfully τ-injective. Indeed, let $A \in T(\tau)$; then any R-linear map $A \to Q_{\tau^e}(Q_\sigma(R))$ extends uniquely to an R-linear $A^e \to Q_{\tau^e}(Q_\sigma(R))$ by property (T) for σ, but this map is $Q_\sigma(R)$-linear by the σ-injectivity of $Q_{\tau^e}(Q_\sigma(R))$. Since $A^e \in T(\tau^e)$, the latter map may be extended to a unique $Q_\sigma(R) \to Q_{\tau^e}(Q_\sigma(R))$ which restricts to $R \to Q_{\tau^e}(Q_\sigma(R))$, providing the desired R-linear map extending the initial $A \to Q_{\tau^e}(Q_\sigma(R))$.

The isomorphism $Q_\tau(Q_\sigma(R)) \cong Q_\tau(R)$ follows from Theorem 10.

THEOREM 47. Let X_A be a geometric open set of $X = \text{Spec } R$, then X_A is an affine scheme, in fact $X_A = \text{Spec } Q_A(R) = X'$.

PROOF. The prime ideals $P \not\supset A$ are exactly the elements $P \notin T(A)$. Proposition 11 implies that these prime ideals are in one-to-one correspondence with proper prime ideals of $Q_A(R)$. An open subset of X_A is of the form X_{AB} hence it is some X_C with $C \subset A$. Because the operations c,e respect inclusions it follows that X_C is in one-to-one correspondence with $X'_{C^e} = \{P' \in X', P' \not\supset C^e\}$ and thus e defines a homeomorphism of the topological spaces X_A and X' (for the induced Zariski topologies).

We associated $Q_C(R)$ to X_C and $Q_{C^e}(Q_A(R))$ to X'_{C^e}. The foregoing lemma yields $Q_{C^e}(Q_A(R)) \cong Q_C(Q_A(R)) \cong Q_C(R)$, proving that X_A with the induced Zariski topology and the restricted sheaf is an affine scheme.

THEOREM 48. Let X_C and X_A be different T-sets in X such that $X_C \subset X_A$ and suppose that X_A is a geometric set. Then X_{C^e} is a T-set in Spec $Q_A(R)$ if and only if X_C is a T-set in Spec R. There is a one-to-one correspondence between T-sets properly contained in X_A and proper T-sets in Spec $Q_A(R)$.

PROOF. First suppose that σ_C is a T-functor. We show that every left ideal $L \in T(C^e)$ is a σ_{C^e}-projective $Q_A(R)$-module and then σ_{C^e} has property (T) because R is left Noetherian. Let $L \in T(C^e)$ and let $M' \to M \to 0$ be an exact sequence of σ_{C^e}-torsion free $Q_A(R)$-modules. Given a $Q_A(R)$-linear map $h : L \to M$. From $L \supset (C^e)^n$ derives $L^c \supset C^n$ or $L^c \in T(C)$. Restriction of h to L^c yields an R-linear $h_c : L^c \to M$. As R-modules, M and M' are σ_C-torsion free and from the fact that σ_C is a T-functor it follows that L^c is σ_C-projective and thus there exists a left ideal $B \in T(C)$, $B \subset L^c$ such that the following diagram of R-linear maps is commutative :

$$\begin{array}{ccccc} B & \hookrightarrow & L^c & & \\ \downarrow{\scriptstyle f'} & & \downarrow{\scriptstyle h_c} & & \\ M' & \longrightarrow & M & \longrightarrow & 0 \end{array}$$

If L is proper then $L^c \notin T(A)$, hence $B \notin T(A)$. Because M is a $Q_A(R)$-module, it is faithfully σ_A-injective (because σ_A is a T-functor). Therefore the $Q_A(R)$-linear map h is defined as follows :
$h(q\,a) = q\,h_c(a)$ with $a \in L^c$, $q \in Q_A(R)$. In a similar way f' extends to a $Q_A(R)$-linear map $f : B^e \to M'$, $f(q\,b) = q\,f'(b)$ for $q \in Q_A(R)$, $b \in B$.

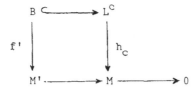

Now, $B \in T(C) - T(A)$, thus $B^e \in T(C^e)$ and $B^e \neq Q_A(R)$; obviously the following diagram commutes :

$$\begin{array}{ccccc} B^e & \hookrightarrow & L & & \\ {\scriptstyle f}\downarrow & & \downarrow{\scriptstyle h} & & \\ M' & \longrightarrow & M & \longrightarrow & 0 \end{array}$$

proving that L is σ_{C^e}-projective. Conversely, assume that σ_{C^e} is a T-functor, let $M \in M(R)$. Since $Q_C(M)$ is a $Q_C(R)$-module, hence a $Q_A(R)$-module via $Q_A(R) \rightarrow Q_C(R)$, it follows that the inclusion $M \rightarrow Q_C(M)$ extends to a $Q_A(R)$-linear $i^* : Q_A(M) \rightarrow Q_C(M)$, (because $Q_C(M)$ is σ_A-faithfully injective). Obviously, $\text{Ker } i^* = Q_A(\sigma_C(M)) \subset \sigma_C(Q_A(M))$. Action of Q_{C^e} on the exact sequence of $Q_A(R)$-modules :

$$0 \rightarrow \text{Ker } i^* \rightarrow Q_A(M) \rightarrow Q_C(M) \rightarrow Q_C(M)/\text{Im } i^* \rightarrow 0$$

yields an exact sequence :

$$0 \rightarrow Q_{C^e}(Q_A(M)) \rightarrow Q_{C^e}(Q_C(M)) \rightarrow 0$$

This is so because $\text{Im } i^* \supset i(M)$ while $Q_C(M)/i(M)$ is σ_C-torsion, thus $Q_C(M)/\text{Im } i^*$ is σ_C-torsion and also σ_{C^e}-torsion as a $Q_A(R)$-module.
The technique of Lemma 46 may be used to proof that $Q_C(M)$ is a faithfully σ_{C^e}-injective $Q_A(R)$-module. Thus $Q_{C^e}(Q_C(M)) \cong Q_C(M)$ and therefore $Q_{C^e}(Q_A(M)) \cong Q_C(M)$ follows.
Now, $Q_{C^e}(Q_A(M)) \cong Q_{C^e}(Q_A(R)) \underset{Q_A(R)}{\otimes} Q_A(M)$ by (T) for σ_{C^e}.
Then, $Q_{C^e}(Q_A(M)) \cong Q_C(R) \underset{Q_A(R)}{\otimes} \left(Q_A(R) \underset{R}{\otimes} M\right) \cong Q_C(R) \underset{R}{\otimes} M$. The isomorphism $Q_C(M) \cong Q_C(R) \underset{R}{\otimes} M$ is equivalent to property (T) for σ_C because R is left Noetherian. Denote C^{ec} by C', i.e., $C' = \{x \in R,\ L x \subset C \text{ for some } L \in T(A)\}$.
The one-to-one correspondence follows easily from the fact that $X_A \cap X_{C'} = X_C$. Indeed, $X_C \subset X_{C'}$ and if $P \in X_{C'} - X_C$ then $P \supset C$, $P \not\supset C'$, but $P \supset L C'$ for some $L \in T(A)$ and thus $P \supset L$, entailing $P \notin X_A$.

<u>PROPOSITION</u> 49. If X_A and X_B are T-sets in Spec R then $X_{AB} = X_A \cap X_B$ is a T-set.

<u>PROOF</u>. We have to prove that $\sigma = \sup\{\sigma_A, \sigma_B\}$ is a T-functor. The filter $T(\sigma)$ has a basis consisting of finite products of ideals in $T(A) \cup T(B)$. Let $C = C_1 \cdot \ldots \cdot C_r$ be such a product. Knowing that $Q_\sigma(R)$ contains both $Q_A(R)$ and $Q_B(R)$ as subrings one easily deduces that $1 \in Q_\sigma(R)C$ from the fact that $Q_\sigma(R)C_i = Q_\sigma(R)$, $i = 1,\ldots,r$. Thus, $C \in T(\sigma)$ yields $Q_\sigma(R)C = Q_\sigma(R)$ and this implies that σ is a T-functor.

<u>Remark</u>. The same argumentation yields that the sup of a set of T-functors is again a T-functor.

<u>PROPOSITION</u> 50. If Spec R has a T-basis then each stalk is a T-stalk.

<u>PROOF</u>. Let X_A be a Zariski open set in X and let $P \in X_A$. The existence of a T-basis entails that X_A is a union of T-sets, hence $P \in X_B \subset X_A$ for some T-set X_B. Since $\sigma_B \geq \sigma_A$, we have that

$$\sigma_{R-P} = \sup\{\sigma_A,\ P \in X_A\} = \sup\{\sigma_A,\ P \in X_A,\ X_A \text{ is a T-set}\}$$

and by the above remark, σ_{R-P} is a T-functor.

<u>THEOREM</u> 51. Let σ_1 and σ_2 be symmetric T-functors, let σ be $\sup\{\sigma_1,\sigma_2\}$. If an ideal A of R is a σ_1- and a σ_2-ideal then A is a σ-ideal.

<u>PROOF</u>. $T(\sigma)$ has a basis of finite products of ideals in $T(\sigma_1) \cup T(\sigma_2)$. We may suppose $A \notin T(\sigma)$, hence $A \notin T(\sigma_1)$ and $A \notin T(\sigma_2)$. Consider $\prod_i C_i \in T(\sigma)$ with $C_i \in T(\sigma_1) \cup T(\sigma_2)$. Then,

$$A C_1 \ldots C_n \supset C_1' A C_2 \ldots C_n \supset \ldots \supset C_1' \ldots C_n' A,$$

with $\prod_i C_i' \in T(\sigma)$; each inclusion deriving from the σ_1- or σ_2-ideal condition for A. Theorem 14 yields that A is a σ-ideal and $Q_\sigma(R)A$ is an ideal of $Q_\sigma(R)$.

COROLLARY 1. If X_A and X_B are geometric sets in Spec R then X_{AB} is a geometric set.

COROLLARY 2. If Spec R has a geometric basis then each stalk $Q_{R-P}(R)$ is a geometric stalk.

For geometric sets, we may prove an analogue of Theorem 48 :

PROPOSITION 52. Let X_C and X_A be different T-sets in Spec R such that $X_C \subset X_A$ and assume that X_A is a geometric set. Then, X_{C^e} is geometric in Spec $Q_A(R)$ if and only if X_C is geometric. Proper geometric subsets of Spec $Q_A(R)$ correspond one-to-one to geometric subsets of X_A.

PROOF. The correspondence of T-sets presents no problem because this has already been proven in Theorem 48. Suppose that X_C is geometric and let B be an ideal of $Q_A(R)$, such that $B \notin T(\sigma_{C^e})$. By contraction, B^c is an ideal of R and it is a σ_A-ideal since $B^{ce} = B$. Now $B^c \notin T(\sigma_C)$ implies that $Q_C(R)B^c$ is an ideal of $Q_C(R)$ and hence, for every $L \in T(\sigma_C)$ there exists an $L' \in T(\sigma_C)$ such that $L'B^c \subset B^cL$. Extension of these ideals to $Q_A(R)$ yields that $(L')^eB \subset B\,L^e$, and since every ideal in $T(\sigma_{C^e})$ is of the form L^e for some $L \in T(\sigma_C)$ it follows that B is σ_{C^e}-ideal. Conversely, suppose that $Q_A(R)$ is σ_{C^e}-perfect and let B be an ideal of R. Then $Q_A(R)B$ is an ideal of $Q_A(R)$ because X_A is geometric. Thus $Q_{C^e}(Q_A(R))B$ is an ideal of $Q_{C^e}(Q_A(R))$. The isomorphism $Q_C(R) \cong Q_{C^e}(Q_A(R))$ finishes the proof. The one-to-one correspondence derives from the foregoing and the fact that $X_{C'} \cap X_A = X_C$ where $C' = \{x \in R,\ L\,x \subset C$ for some $L \in T(\sigma_A)\}$.

It has been pointed out before that it is a most interesting problem to determine a class of rings which are σ_A-perfect for every ideal A. For such a ring R it follows that every T-set X_A in Spec R is geometric, and so this brings us closer to the commutative case.

A ring R satisfies the Artin-Rees condition if and only if for any pair of ideals A,B of R, there exists an integer $n > 0$ such that $B \cap A^n \subset BA$. Because of its obvious consequences for the theory of reductions presented in section II. 3., we include the following proposition in its most general form.

PROPOSITION 53. Let R be a left Noetherian ring satisfying the Artin-Rees condition and let σ be a symmetric T-functor. Then :

1. $Q_\sigma(R/A)$ is an R/A-module for every ideal A of R.
2. The R/A-module structure of $Q_\sigma(R/A)$ extends uniquely to a ring structure and the R-linear $\pi_\sigma : Q_\sigma(R) \to Q_\sigma(R/A)$ induced by the canonical map $\pi : R \to R/A$ is a ring homomorphism.

PROOF. 1. It is sufficient to show that in the R-module structure A annihilates $Q_\sigma(R/A)$. Elements of $Q_\sigma(R/A)$ may be represented as [C,f] with $C \in T(\sigma)$, $f \in \mathrm{Hom}_R(C, R/A)$ and $[C,f] = [C',f']$ if and only if f and f' coincide on a left ideal $B \in T(\sigma)$ such that $B \subset C \cap C'$. The R-module structure is then given by $x[C,f] = [C',g]$ where $C' \in T(\sigma)$ satisfies $C'x \subset C$ and $g(c') = f(c'x)$ for $c' \in C'$, $x \in R$. Take n such that $C^n \cap A \subset AC$. Since $C^n \in T(\sigma)$ we have that

$$x[C,f] = x[C^n, f|C^n] = [C',g],$$

where $C'x \subset C^n$. If $x \in A$, $C'x \subset C^n \cap A \subset AC$ and thus for $c' \in C'$, $g(c') = f(C'x) = 0$ because $C'x \in AC$.

2. Proceeding as in Proposition 33, this follows easily. Since π_σ is also induced by $R/\sigma(R) \to R/A_\sigma$, another proof for 2. may be given if one

modifies 1. as follows; $Q_\sigma(R/A)$ is an R/A_σ-module. Then Proposition 33 applies directly because R/A_σ is σ-torsion free.

References for Section III.

D.C. MURDOCH, F. VAN OYSTAEYEN [26], F. VAN OYSTAEYEN [38].

IV. PRIMES IN ALGEBRAS OVER FIELDS

IV. 1. Pseudo-places of Algebras over Fields.

Let K be any field and let A be a K-ring. A pseudo-place of the K-ring A is given by a triple $(A',\psi,A_1/K_1)$, where $\psi : A' \to A_1$ is a ring homomorphism defined on a subring A' of A such that $A' \cap K$ is a valuation ring O_K of K, and such that the restriction of ψ to $A' \cap K$ is a place of K whith residue field K_1. In the sequel, pseudo-places and places will always be assumed to be surjective, and unless otherwise specified, K and K_1 will be contained in the center of A and A_1 respectively. A pseudo-place $(A',\psi,A_1/K_1)$ of the K-algebra A will be denoted by ψ when no confusion is possible. If ψ is a pseudo-place of A/K, and if $\{y_1,\ldots,y_m\}$ is K_1-independent in A then any set of representatives $\{x_1,\ldots,x_m\} \subset A'$ with $\psi(x_i) = y_i$ is clearly K-independent in A.

Examples of pseudo-places are : places of fields, homomorphisms of algebras over fields and specializations of orders in central simple algebras. The substantial part of A with respect to $(A',\psi,A_1/K_1)$ is the subalgebra $K[A']$ generated over K by the elements of A'. A pseudo-place ψ is special if for every $x \in A$ there is a $\lambda \in O_K$, $\lambda \neq 0$, such that $\lambda x \in A'$, i.e. if and only if $A = K[A']$. It is in general not really restricting to consider special pseudo-places only, however we do not systematically impose this in the sequel.

A pseudo-place ψ of A/K is said to be restricted if for every non-zero $x \in A$ there is a $\lambda \in K$ such that $\lambda x \in A' - P$, where $P = \operatorname{Ker}\psi$. If $[A : K] < \infty$, then ψ is unramified if $[A_1 : K_1] = [A : K]$. The properties unramified, restricted, special, decrease in strength if listed in this order.

PROPOSITION 54. Let $n = [A : K] < \infty$ and let ϕ be a place of K with valuation ring O_K and maximal ideal M_K. Then there exists at least one unramified pseudo-place ψ of A such that the restriction of ψ to K is isomorphic to ϕ.

PROOF. Let $\{c_{ij,k} \mid i,j,k = 1,\ldots,n\} \subset K$ be a set of structural constants associated to a K-basis $\{e_1,\ldots,e_n\}$ of A. There is an $a \in O_K$ such that $a\, c_{ij,k} = c^*_{ij,k} \in O_K$ for all i,j,k. Putting $f_i = a\, e_i$, we obtain a K-basis $\{f_1,\ldots,f_n\}$ of A such that the corresponding set of structural constants $\{c^*_{ij,k}\}$ is a subset of O_K. The O_K-module A' generated by $\{f_1,\ldots,f_n\}$ is then obviously a ring and the M_K-module M generated by $\{f_1,\ldots,f_n\}$ is an ideal in A'. The canonical epimorphism $\psi : A' \to A'/M = A_1$ defines a pseudo-place of A/K. Indeed, observe that if $\sum_{i=1}^n \alpha_i f_i = \lambda \in K - O_K$ with $\alpha_i \in O_K$, then because $\lambda^{-1}\alpha_i \in M_K$ it follows that $1 \in M$. Hence $1 = \sum_{i=1}^n m_i f_i$, $m_i \in M_K$. Multiplying f_j by 1 on the left yields : $f_j = \sum_{k,i=1}^n m_i\, c^*_{ij,k}\, f_k$, and we obtain $1 = \sum_{i=1}^n m_i\, c^*_{ij,j} \in M_K$, a contradiction. Moreover, $\delta = \sum_{i=1}^n m_i f_i \in O_K - M_K$, with $m_i \in M_K$ entails $1 = \sum_{i=1}^n (m_i \delta^{-1}) f_i \in M$, again a contradiction, hence $M \cap K = M_K$. To prove that ψ is unramified, let $\sum_{i=1}^n \mu_i\, \psi(f_i) = 0$ with $\mu_i \in K_1$. Choose representatives $\lambda_i \in O_K$ such that $\phi(\lambda_i) = \mu_i$, $i = 1,\ldots,n$. Then

$$\psi\Big(\sum_{i=1}^n \lambda_i f_i\Big) = 0 \quad \text{or} \quad \sum_{i=1}^n \lambda_i f_i \in M,$$

entailing $\lambda_i \in M_K$ and $\mu_i = 0$.

DEFINITION. Let k be a subfield of K such that the restriction of the pseudo-place ψ to K is a k-place of K then ψ is called a k-pseudo-place of A. The dimension of a k-pseudo-place ψ is defined to be the transcendence degree of K_1 over k. The rank of an unramified pseudo-place ψ of A/K is the rank of the place ψ/K. Let B be a K-algebra, $K \subset B \subset A$. The restriction of a pseudo-place ψ of A/K to B is a pseudo-place of B/K which is special, restricted, if ψ is such.

PROPOSITION 55. The restriction of an unramified pseudo-place ψ of A/K to subalgebra B/K is an unramified pseudo-place of B/K.

PROOF. Let $\{y_1,\dots,y_r\}$ be a K_1-basis of $B_1 = \psi(B \cap A')$ and complete it to a K_1-basis of A_1, $\{y_1,\dots,y_{r+1},\dots,y_n\}$. Choose a set of representatives $\{x_1,\dots,x_n\}$ with $\psi(x_i) = y_i$ and with $\{x_1,\dots,x_r\} \subset B$. The set $\{x_1,\dots,x_n\}$ is a K-basis of A and if $\{x_1,\dots,x_r\}$ is not a K-basis of B then it may be completed to one, $\{x_1,\dots,x_r,b_{r+1},\dots,b_q\}$ say. We write down the $q-r$ relations : $b_{r+t} = \sum_{j=1}^{r} a_{tj}\, x_j + \sum_{k=r+1}^{n} c_{tk}\, x_k$ with a_{tj} and c_{tk} in K.

Put $b'_{r+t} = \sum_{k=r+1}^{n} c_{tk}\, x_k$. Then $\{x_1,\dots,x_r,b'_{r+1},\dots,b'_q\}$ is still a K-basis of B. Fix an index t. There exists a $c_t \in K$ such that $c_t c_{tk} \in O_K$ for all $r+1 \leq k \leq n$ and $c_t c_{tl} = 1$ for some $r+1 \leq l \leq n$. We obtain $b^*_{r+t} = c_t b'_{r+t} \in B \cap A' = B'$ and $\psi(b^*_{r+t}) = \sum_{k=r+1}^{n} \psi(c_t c_{tk}) y_k \in \psi(B') = B_1$.
For suitable $\xi_i \in K_1$ $i = 1,\dots,r$, we have $\sum_{k=r+1}^{n} \psi(c_t c_{tk}) y_k = \sum_{i=1}^{n} \xi_i y_i$, with $\psi(c_t c_{tl}) = 1$, hence a contradiction.

The following theorem characterizes unramified pseudo-places; it is fundamental for the application of pseudo-places to the theory of central simple algebras (section V).

THEOREM 56. Let ψ be an unramified pseudo-place of A/K then :

1. A' is a free O_K-module of rank $n = [A : K]$.
2. There exists a K-basis $\{e_1,\dots,e_n\}$ of A, generating A' over O_K and such that $\{\psi(e_1),\dots,\psi(e_n)\}$ is a K_1-basis of A_1.

PROOF. Pick a K-basis $E = \{e_1,\dots,e_n\}$ of A such that $e_i \in A'$ for all $i = 1,\dots,n$, and such that $\psi(E)$ is a K_1-basis of A_1. The O_K-module generated by E is a ring; for, let $e_i e_j = \sum_{k=1}^{n} c_{ij,k}\, e_k$, there exist $c_{ij} \in K$ such that $c_{ij} c_{ij,k} = c^*_{ij,k}$ is in O_K for all $k = 1\dots n$, while $c^*_{ij,l} = 1$ for some l. Hence, $c_{ij} e_i e_j = \sum_{k=1}^{n} c^*_{ij,k} e_k \in A'$.

For a couple (i,j) such that $c_{ij} \notin M_K$ we have $c_{ij}^{-1} \in O_K$ hence $c_{ij,k} = c_{ij}^{-1}c_{ij,k}^{*} \in O_K$. For a couple (i,j) with $c_{ij} \in M_K$ we get on the one hand : $\psi(c_{ij}e_ie_j) = 0$ and on the other hand

$$\psi(c_{ij}e_ie_j) = \Sigma_{k=1}^{n} \psi(c_{ij,k}^{*})\ \psi(e_k),$$

contradicting the K_1-linear idependency of $\psi(E)$. So for every couple (i,j) the product e_ie_j belongs to the O_K-module generated by E. Pick any $x \in A'$, $x = \Sigma_{i=1}^{n} x_ie_i$ with $x_i \in K$. For some index j, $1 \leqslant j \leqslant n$, we have $x_j^{-1}x = \Sigma_{i=1}^{n} x_i^{*} e_i$ with $x_i^{*} \in O_K$, $x_j^{*} = 1$. Now, $x_j^{-1} \notin M_K$ implies that x_j is in O_K and thus all x_i are in O_K too, meaning that x belongs to $O_K[E]$. Furthermore, $x_j^{-1} \in M_K$ yields a contradiction because then a non-trivial relation $0 = \psi(x_j^{-1})\psi(x) = \Sigma_{i=1}^{n} \psi(x_i^{*})\psi(e_i)$, follows. Thus $A' = O_K[E]$ and the proof is complete.

If L is a division ring containing the field K in its center and such that $[L : K] < \infty$ then an unramified pseudo-place $(L', \psi, L_1/K_1)$ of L/K is a place of L if and only if L_1 is a division ring. The definition of a place of a skew-field is analogous to the corresponding definition in the commutative case, (cf. [34]). We wish to extend this definition as follows.

A <u>pre-place</u> of a K-algebra A is a pseudo-place ψ such that :

P_1 : $\psi(A') = A_1$ is a division ring.

P_2 : If $x,y \in A$ and $xy \in A'$ then $x \notin A'$ implies $y \in P$, similarly, $y \notin A'$ implies $x \in P$, where $P = \mathrm{Ker}\ \psi$.

We use the term division ring to refer to a ring without zero divisors, i.e. a ring such that (0) is a completely prime ideal.

<u>PROPOSITION</u> 57. Let $(A',\psi,A_1/K_1)$ be a restricted pseudo-place such that A_1 is a division ring, then ψ is a pre-place.

PROOF. Observe that if $x \notin A'$ and if $\lambda \in K$ is such that $\lambda x \in A' - P$ then $\lambda \in M_K$. Thus if $xy \in A'$ with $x \notin A'$ and $y \notin P$, then there exist $\lambda \in M_K$, $\mu \in O_K$ so that $\lambda x \in A' - P$, $\mu y \in A' - P$. Hence $\lambda\mu xy \in A' - P$, but $\lambda\mu \in M_K$ entails $\psi(\lambda x)\psi(\mu y) = 0$ and this contradicts $\lambda x \mu y \in A' - P$. Similarly, $xy \in A'$, $y \notin A'$ implies $x \in P$.

The following shows that restricted pre-places do generalize places of fields.

PROPOSITION 58. Let L be a field, $K \subset L \subset A$. A restricted pre-place ψ of A/K restricts to a place of L.

PROOF. Let $x \in L$ and suppose that $x \notin A' \cap L$ and $x^{-1} \notin A'$. Take $\lambda,\mu \in M_K$ such that μx^{-1} and λx are in $(A' \cap P) \cap L$. Since this set is multiplicatively closed, $\lambda\mu \in (A' - P) \cap L$ follows, but this contradicts $\lambda\mu \in M_K$. Moreover $xx^{-1} \in A'$ with $x \notin A'$ entails $x^{-1} \in L \cap P$. Thus $P \cap L$ is the ideal of non-invertible elements in $A' \cap L$, and therefore $A' \cap L$ is a valuation ring of L.

It is easily verified that an unramified k-pseudo-place of a K-algebra A such that $[A : K] < \infty$ is an isomorphism of A. This follows from Theorem 56 and the fact that a k-place of the finite dimensional field extension K/k is an isomorphism of K.

IV. 2. Specialization of Pseudo-places.

Let $(A_1',\psi_1,A_1/K_1)$ and $(A_2',\psi_2,A_2/K_2)$ be pseudo-places of A/K. We say that ψ_2 is a specialization of ψ_1, and write $\psi_1 \to \psi_2$, if and only if $A_1' \supset A_2'$ and $\operatorname{Ker}\psi_1 \subset \operatorname{Ker}\psi_2$. Obviously, the given definition implies that the place $\psi_1|K$ is a specialization (of places) of $\psi_2|K$.

PROPOSITION 59. Let $(A',\psi,A_1/K_1)$ be a pseudo-place of A/K and let $(A_1',\Omega,A_2/K_2)$ be a pseudo-place of A_1/K_1 then $\Omega \circ \psi$ defines a pseudo-place of A/K on the subring $\psi^{-1}(A_1')$.

PROOF. The only thing to establish is that $\psi^{-1}(A_1') \cap K$ is a valuation ring of K which coincides with $\psi^{-1}(K_1 \cap A_1') \cap K$. It is immediate that $O_K = A' \cap K$ contains $\psi^{-1}(K_1 \cap A_1') \cap K$. Now let $x \in K - \psi^{-1}(A_1' \cap K_1)$, then we distinguish three cases :

case 1. $x \notin O_K$. Then $x^{-1} \in M_K$, thus $\psi(x^{-1}) = 0$ and this yields $x^{-1} \in \psi^{-1}(A_1' \cap K_1) \cap K$.

case 2. $x \in O_K - M_K$. Then $x^{-1} \in O_K$ and $\psi(x^{-1}) \in K_1$. Since $\psi(x) \notin A_1' \cap K_1$ it follows that $\psi(x^{-1}) \in A_1' \cap K_1$, therefore $x^{-1} \in \psi^{-1}(A_1' \cap K_1)$.

case 3. $x \in M_K$. From $\psi(x) = 0$ immediately $x \in \psi^{-1}(A_1' \cap K_1) \cap K$.

Thus, $\psi^{-1}(A_1' \cap K_1) \cap K$ is indeed a valuation ring of K. That it coincides with $\psi^{-1}(A_1') \cap K$ is obvious.

PROPOSITION 60. Let $(A_1',\psi_1,A_1/K_1)$ and $(A_2',\psi_2,A_2/K_2)$ be pseudo-places of A/K. Then $\psi_1 \to \psi_2$ if and only if there exists a pseudo-place Ω of A_1/K_1 such that $\psi_2 = \Omega \circ \psi_1$.

The proof is straightforward and consists in showing that $\psi_1(A_2') \to \psi_1(A_2')/\psi_1(\mathrm{Ker}\ \psi_2) \cong A_2$ is the desired pseudo-place Ω.

Remark. ψ_2 is unramified if and only if ψ_1 and Ω are unramified. The same is true for restricted or special pseudo-places.

Furthermore, in the situation described in Proposition 60, there exists a K-basis $\{u_1,\dots,u_n\}$ of A such that : $A_1' = O_1[u_1,\dots,u_n]$ with $O_1 = A_1' \cap K$, and $A_2' = O_2[u_1,\dots,u_n]$ with $O_2 = A_2' \cap K$.
This allows us to construct, for any algebra A/K, for arbitrary places ϕ_1 and ϕ_2 of K such that $\phi_1 \to \phi_2$ and for any unramified pseudo-place $(A_2',\psi_2,A_2/K_2)$ of A/K such that $\psi_2/K = \phi_2$, an unramified pseudo-place $(A_1',\psi_1,A_1/K_1)$ of A/K such that $\psi_1 \to \psi_2$ and $\psi_1/K = \phi_1$.

With notations as above; pseudo-places ψ_1 and ψ_2 are said to be <u>isomorphic</u> if and only if there exists an isomorphism $\Omega : A_1 \to A_2$ such that $\Omega(K_1) = K_2$ and $\psi_2 = \Omega \circ \psi_1$. This definition implies that if $\psi_1 \cong \psi_2$ then $\psi_1 \to \psi_2$ and $\psi_2 \to \psi_1$, hence both ψ_1 and ψ_2 are defined on the same subring of A and their kernels coincide.
If $\psi_1 \to \psi_2$ are unramified pseudo-places such that $\psi_1/K \cong \psi_2/K$ (as places) then $\psi_1 \cong \psi_2$. Moreover, if $\psi_1 \to \psi_2$ are unramified k-pseudo-places of A/K such that $\dim \psi_1/k = \dim \psi_2/k$ is finite then ψ_1 is isomorphic to ψ_2.

Let $(A',\psi,A_1/K_1)$ be an unramified pseudo-place of finite rank m, i.e. the place $\phi = \psi/K$ has finite rank m. For such places there exists a specialization chain : $\phi_{m-1} \to \phi_{m-2} \to \ldots \to \phi_1 \to \phi$ (*), associating a place of rank 1 to the given ϕ. The observations made above, imply that we can find a specialization chain of non-isomorphic unramified pseudo-places of A/K : $\psi_{m-1} \to \ldots \to \psi_1 \to \psi$ (**), with $\psi_i/K = \phi_i$. Such a chain is said to be <u>maximal</u> if it is impossible to insert an unramified pseudo-place ξ of A/K such that $\psi_{m-i} \to \xi \to \psi_{m-i-1}$, and ξ not being isomorphic to ψ_{m-i} or ψ_{m-i-1}. Maximal chains of this type do exist and moreover, a chain (**) is maximal if and only if the associated chain of places (*) is a maximal one; this justifies our definition of the rank of an unramified pseudo-place.
Amongst further properties of pseudo-places, we mention that pseudo-places are "reduceable" under surjective ring homomorphisms. Let A be a K-algebra, J and ideal of A, and let $(A',\psi,A_1/K_1)$ be a pseudo-place of A/K. There exists a pseudo-place Ω of A/J such that the following diagram of pseudo-places is commutative, i.e. $\Omega\pi = \pi_1\psi$ on A' :

$$\begin{array}{ccc} A & \xrightarrow{\pi} & A/J \\ \downarrow \psi & & \downarrow \Omega \\ A_1 & \xrightarrow{\pi_1} & A_1/J_1 \end{array}$$

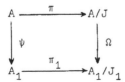

where $J_1 = \psi(J \cap A')$. Moreover, if ψ is unramified then Ω is unramified too, (proofs in [34]). Note also that if A,B are K-algebras, and if ψ,Ω are pseudo-places of A,B respectively, inducing the same place ϕ of K then the tensorproduct $\psi \underset{\phi}{\otimes} \Omega$ is defined to be the pseudo-place

$$(A' \underset{O_K}{\otimes} B' , \psi \otimes \Omega, A_1 \underset{K_1}{\otimes} B_1) \quad \text{of} \quad A \underset{K}{\otimes} B.$$

We now focus on pre-places.

PROPOSITION 61. Let $(A',\psi,A_1/K_1)$ be a pre-place of A/K and let $(A_1',\Omega,A_2/K_2)$ be a pre-place of A_1/K_1 then $(\psi^{-1}(A_1'), \Omega \circ \psi, A_2/K_2)$ is a pre-place of A/K.

PROOF. Since A_2 is a division ring and because $\Omega \circ \psi$ defines a pseudo-place of A/K it will be sufficient to verify the condition P_2. Suppose that $xy \in \psi^{-1}(A_1')$ while $x \notin \psi^{-1}(A_1')$, then :

1. Both x and y are in A'. Then $\psi(x)\psi(y) = \psi(xy) \in A_1'$ and $\psi(x) \notin A_1'$ yield $\psi(y) \in \operatorname{Ker} \Omega$, thus $y \in \operatorname{Ker} \Omega \circ \psi$.
2. If x or y is not in A', then $xy \in \psi^{-1}(A_1') \subset A'$ implies that y or x resp. is in $\operatorname{Ker} \psi \subset \operatorname{Ker} \Omega \circ \psi$.

PROPOSITION 62. Let ψ_1 and ψ_2 be pre-places of A/K then equivalently :

1. ψ_2 is a specialization of ψ_1.
2. There is a pre-place Ω on the residue division algebra of ψ_1 such that $\Omega \circ \psi_1 = \psi_2$.

PROOF. The implication 2 ⇒ 1 is obvious. Conversely, Proposition 60 implies that a pseudo-place Ω of the residue division algebra of ψ_1 exists such that $\Omega \circ \psi_1 = \psi_2$. The residue algebra of ψ_2 is also a division ring, so we are left to verify the condition P_2.
Let $(A_1',\psi_1,A_1/K_1)$ and $(A_2',\psi_2,A_2/K_2)$ be the pre-places. Since ψ_1

specializes to ψ_2 we get $A_1' \supset A_2' \supset \text{Ker } \psi_2 \supset \text{Ker } \psi_1$ and thus $\psi_1^{-1}\psi_1(A_2') = A_2'$. Write $B = \psi_1(A_2')$. If $\bar{x}\bar{y} \in B$ with $\bar{x},\bar{y} \in A_1$ and with $\bar{x} \notin B$, then for representatives x,y of $\bar{x},\bar{y}$ resp. we obtain : $x y \in A_2'$ with $x,y \in A_1'$ but $x \notin A_2'$. Thus, $y \in \text{Ker } \psi_2$ and $\bar{y} \in \psi_1(\text{Ker } \psi_2)$ or $\bar{y} \in \text{Ker } \Omega$ follows. Similarly, $\bar{x}\bar{y} \in B$ with $\bar{y} \notin B$ implies that $\bar{x} \in \psi_1(\text{Ker } \psi_2)$.
These results may be restated as follows.
Let $(A',\psi,A_1/K_1)$ be a pseudo-place of A/K. Write $PS_K(A)$ for the set of pseudo-places of A/K, then ψ induces a map $\tilde{\psi} : PS_{K_1}(A_1) \to PS_K(A)$ defined by $\tilde{\psi}(\Omega) = \Omega \circ \psi$. Observe that Im $\tilde{\psi}$ consists of all specializations of ψ. Obviously, Im $\tilde{\psi}$ contains unramified pseudo-places if and only if ψ is unramified an then $\tilde{\psi}$ restricts to a map $UP_{K_1}(A_1) \to UP_K(A)$, where $UP_K(A)$ stands for the unramified pseudo-places of A/K. (Similar statements hold for restricted and special pseudo-places). If ψ is a pre-place of A/K, then $\tilde{\psi}$ restricts to $\tilde{\psi} : PR_{K_1}(A_1) \to PR_K(A)$, where $PR_K(A)$ denotes the set of pre-places of A/K. Obviously, in general $\tilde{\psi}$ is injective, also the restrictions of $\tilde{\psi}$ are injective. Reformulating Proposition 62 yields :

$$\tilde{\psi}(PR_{K_1}(A_1)) = PR_K(A) \cap \text{Im } \tilde{\psi}.$$

It is clear that elements of $PS_{K_1}(A_1)$, (or $PR_{K_1}(A_1)$), are isomorphic if and only if their images under $\tilde{\psi}$ are isomorphic in $PS_K(A)$, (or $PR_K(A)$), so $\tilde{\psi}$ induces an injective map on the isomorphism classes of pre-places. This entails that $\tilde{\psi}(\Omega)$ is, up to isomorphism, only depending on A_1' and Ker Ω, this will be studied in IV. 4.

Remark. In general it may be quite difficult to construct pseudo-places ψ such that Im $\tilde{\psi}$ contains a prescribed set $S \subset PS_K(A)$.

Given a field k and a k-algebra D. If a field K and a K-algebra $\mathcal{D}$ can be constructed such that a pseudo-place ψ of $\mathcal{D}$/K with residue algebra D/k exists, for which Im $\tilde{\psi}$ contains a prescribed subset $S \subset PS_K(\mathcal{D})$ then $\mathcal{D}$/K is said to be S-generic for D/k.
Examples of generic constructions are given in Section V.

Because a surjective K-algebra morphism is a pseudo-place (and a pre-place if its Ker is completely prime) the above statements hold for surjective K-algebra morphisms. However, we want to get rid of the surjectivity hypothesis.

PROPOSITION 63. To a K-algebra morphism $f : A \to B$ there corresponds a mapping $\tilde{f} : PR_K(B) \to PR_K(A)$, defined by $\tilde{f}(\psi) = \psi \circ f$.

PROOF. Let $(B', \psi, B_1/K_1)$ be a pre-place of B/K. Put $A' = f^{-1}(B')$, a pseudo-place of A/K may be defined by $\psi \circ f : A' \to \psi(ff^{-1}(B')) \subset B_1$. Now, as a subring of B_1, $\psi(ff^{-1}(B'))$ is a division ring, thus only the condition P_2 has to be checked. Suppose $x, y \in A$ are given, such that $xy \in A'$ with $x \notin A'$. Then $f(xy) = f(x)f(y) \in B'$, but $f(x) \notin B'$ yields $f(y) \in \text{Ker}\,\psi$ or $y \in f^{-1}(\text{Ker}\,\psi)$. In a similar way it may be shown that, if $xy \in A'$ and $y \in A'$, then $x \in f^{-1}(\text{Ker}\,\psi)$, proving that $\psi \circ f$ is a pre-place of A.

Let $\underline{Alg}_K$ be the category of K-algebras with K-algebra morphisms (this may be generalized to a category of algebras using pseudo-places or pre-places for the morphisms). It is possible to put topologies on the sets $PS_K(A)$ such that we get a functor $PS_K : \underline{Alg}_K \to \underline{Top}$.

An interesting variation on this theme is subject of IV. 4., additional interest is added there by the existence of a fitting localization technique.

Remark. If f is not surjective, $\tilde{f}$ is not necessarily injective. Conditions for $\tilde{f}$ to be injective may be expressed in terms of the algebra-extension B of f(A).

IV. 3. Pseudo-places of Simple Algebras.

PROPOSITION 64. Let A be a K-algebra, let $(A', \psi_1, A_1/K_1) \in UP_K(A)$ be so that A_1 is a K_1-central simple algebra. Then A is a K-central simple algebra.

PROOF. Suppose that A is not simple and let J be a proper ideal of A. Then $I = J \cap A'$ is an ideal of A', and $\psi(I)$ is an ideal in A_1. There are two possibilities left :

1. $\psi(I) = A_1$. Then I contains a K-basis for A and thus J = A.
2. $\psi(I) = 0$. Let $A' = 0_K[E]$, where E is a K-basis for A as in Theorem 56. If a non-zero $x \in I$ exists, then $x = \Sigma_{i=1}^n a_i e_i$ with $a_i \in 0_K$, $e_i \in E$. Now, $\psi(x) = 0$ entails $a_i \in M_K$ for all $i = 1,\ldots,n$, and for an index j, $1 \leqslant j \leqslant n$, $a_j^{-1}x = \Sigma_{i=1}^n a_i^* e_i$ with $a_i^* \in 0_K$ and $a_j^* = 1$. Hence $a_j^{-1}x \in A' \cap J$ because J is an ideal. This means that $\psi(a_j^{-1}x) = 0$, but $0 = \Sigma_{i=1}^n \psi(a_i^*)\psi(e_i)$ is contradicting the choice of the K-basis E; I = 0 follows. Since every non-zero element of J may be multiplied by an element in K to yield an element in I it follows that J = (0).

Finally, ψ restricts to an unramified pseudo-place of the center Z(A) of A (Proposition 55), and since the residue algebra of Z(A) is in the center of A_1, the equality $[A : K] = [A_1 : K_1]$ yields that Z(A) = K.

PROPOSITION 65. Let A be a finite dimensional K-algebra, let $\psi \in UP_K(A)$.

1. If A_1 is semi-simple then A is semi-simple.
2. If A_1 is a K_1-central division algebra then A is a K-central division algebra.

PROOF. The first statement is a consequence of the "descent" of nilpotent ideals of A to nilpotent ideals of A_1. Secondly, Proposition 64 yields that A is a K-central simple algebra and it is straightforward to check that zero-divisors in A reduce to zero-divisors in A_1, cf. [34].

Let L/K be a field extension with Galois group G. A <u>crossed product</u> algebra $A = (G, L/K, \{C_{\sigma,\tau}\})$ is defined to be the algebra $L[U_\sigma, \sigma \in G]$ where U_σ, $\sigma \in G$, are symbols satisfying the relations $U_\sigma U_\tau = C_{\sigma,\tau} U_{\sigma\tau}$, $\sigma,\tau \in G$. The set $\{C_{\sigma,\tau}\}$ defines a 2-cocycle, $[\{C_{\sigma,\tau}\}] \in H_2(G,L^*)$. Crossed product algebras are K-central simple. If $A = (G,L/K,\{C_{\sigma,\tau}\})$ is a crossed product algebra, then $\psi \in UP_K(A)$ will be called a <u>galoisian</u> pseudo-place if :

G_1 : $\psi|L \in UP_K(L)$, $G = Gal(L_1/K_1)$ and ψ is compatible with Galois-action in L and L_1.

G_2 : $A_1 = (G,L_1/K_1,\{\overline{C}_{\sigma,\tau}\}$ where $\overline{C}_{\sigma,\tau} = \psi(C_{\sigma,\tau})$ for all $\sigma,\tau \in G$.

With these notations and conventions :

<u>PROPOSITION</u> 66. Let ψ be a galoisian pseudo-place. If e,e_1 are the exponents of the crossed products A,A_1 in the respective Brauer groups $\mathcal{B}r(K)$, $\mathcal{B}r(K_1)$, then e is a multiple of e_1.

<u>PROOF</u>. In $\mathcal{B}r(K)$ we have $A^e = 1$, this means that we can find elements $f_\sigma \in L$, for every $\sigma \in G$, such that, $C^e_{\sigma,\tau} = f_\sigma f^\sigma_\tau f^{-1}_{\sigma\tau}$, $\sigma,\tau \in G$ (*). There exists an $\alpha \in K$ such that $\alpha f_\sigma \in O_L = A' \cap L$ for every $\sigma \in G$ and $\alpha f_\gamma \notin M_L$ for some $\gamma \in G$. Suppose $\alpha \in M_K$. Then for the particular $\gamma \in G$ we derive from (*) a relation :

$$\alpha\, C^e_{\gamma,\gamma}\, \alpha f_{\gamma^2} = \alpha f_\gamma (\alpha f_\gamma)^\gamma .$$

Taking images under ψ yields a contradiction :

$$\psi(\alpha f_\gamma)\psi(\alpha f_\gamma)^\gamma = \psi(\alpha)\psi(C^e_{\gamma,\gamma})\psi(\alpha f_{\gamma^2}) = 0.$$

Hence $\alpha \notin M_K$ and $\alpha^{-1} \in O_K$ follows. Thus $f_\sigma \in O_L$ for all $\sigma \in G$. Moreover, if σ is any element of G then there may be found an element $\tau_\sigma \in G$ such that $\sigma\tau_\sigma = \gamma$, γ the fixed element as before. This yields :

$$C^e_{\sigma,\tau_\sigma} = f_\sigma f^\sigma_{\tau_\sigma} f^{-1}_\gamma ,$$

$$\text{or} \quad \alpha f_\gamma \, C^e_{\sigma,\tau_\sigma} = \alpha f_\sigma f^\sigma_{\tau_\sigma} .$$

Since the left hand side is not in Ker ψ, also $\alpha f_\sigma \notin$ Ker ψ and $f^\sigma_{\tau_\sigma} \notin$ Ker ψ. Hence $f_{\tau_\sigma} \notin M_L$. Now, γ being fixed, if σ runs through G then τ_σ runs through G, hence $f_\sigma \in O_L - M_L$ for every $\sigma \in G$. Consequently, $\overline{C}^e_{\sigma,\tau} = \psi(C^e_{\sigma,\tau}) = \psi(f_\sigma)\psi(f^\sigma_\tau)\psi(f^{-1}_{\sigma\tau})$ and thus $\overline{C}^e_{\sigma,\tau} = \psi(f_\sigma)\psi(f_\tau)^\sigma\psi(f_{\sigma\tau})^{-1}$, implying that $\{\overline{C}^e_{\sigma,\tau}\} \sim 1$. The exponent e_1 being the smallest integer such that $\{\overline{C}^{e_1}_{\sigma,\tau}\}$ is equivalent to the trivial factor set 1, it follows that e_1 divides e.

A pseudo-place ψ of a K-algebra A such that ψ/K is a place of K associated to a Noetherian valuation ring is said to be a Noetherian pseudo-place. There is an obvious link between Noetherian pseudo-places on K-central simple algebras and orders in these algebras. When K is moreover complete with respect to the valuation corresponding to O_K and supposing that ψ is an unramified pseudo-place such that the defining ring A' is a maximal order in A then A_1 is simple (a skew-field) if and only if A is simple (a skew-field). Noetherian pseudo-places also relate to the theory of separable algebras, cf. [2], [3]. We mention that, in case ψ is an unramified Noetherian pseudo-place of A/K such that A_1/K_1 is a separable algebra, then :

1. A is a K-central simple algebra.
2. The residue algebras of unramified pseudo-places defined on maximal orders of A are separable K_1-algebras.
3. The subring A' of A where ψ is defined is itself a maximal order of A.

Similar assertions are still true for non-Noetherian unramified pseudo-places. See Section VI on Azumaya algebras.

IV. 4. Primes in Algebras over Fields.

Let A be a K-algebra and let ψ be a pre-place of A/K. We say that $P = \operatorname{Ker} \psi$ is a prime of A/K and ψ is said to represent the prime P. Note that a prime may be represented by several different pre-places. If ψ represents the prime P, then P is called a ϕ-prime of A/K, where ϕ is the place of K induced by ψ. A ϕ-prime is said to be symmetric if for every $y \in A$, $yP \subset P$ is equivalent to $Py \subset P$. Places of K are considered up to isomorphism unless otherwise specified. We agree to the following notations :

$$\phi - \operatorname{Prim}_K(A) = \{P,\ P \text{ is a } \phi\text{-prime of } A/K\}$$
$$\operatorname{Prim}_K(A) = \{P,\ P \text{ is a prime of } A/K\}.$$

Since the place ϕ corresponding to P is up to isomorphism determined by P we obtain that : $\operatorname{Prim}_K(A) = \coprod_{\phi \in \rho} \phi\text{-}\operatorname{Prim}_K(A)$, with ρ being the set of isomorphism classes of places of K, i.e., the set of valuation rings of K. We introduce a Zariski topology in $\operatorname{Prim}_K(A)$ and the topology induced in $\phi\text{-}\operatorname{Prim}_K(A)$ is called the ϕ-topology. Let F be any finite subset of A, and let $D(F) = \{P \text{ prime of } A/K,\ P \cap F = \phi\}$. The sets $D(F)$, F finite subset of A, form a basis of the topology generated by them because $D(F_1 \cup F_2) = D(F_1) \cap D(F_2)$.

PROPOSITION 67. Prim_K and $\phi\text{-}\operatorname{Prim}_K$ are contravariant functors $\underline{\mathrm{Alg}_K} \to \underline{\mathrm{Top}}$.

PROOF. Given a K-algebra morphism $f : A \to B$. Suppose $(B', \Omega, B_1/K_1)$ is a pre-place of B/K representing a ϕ-prime Q of B/K. Then, $P = f^{-1}(Q) \subset f^{-1}(B')$ defines a ϕ-pseudo-place of A/K with residue algebra $f^{-1}(B')/P$, which is a division ring because it is a subring of B_1. If $x,y \in A$ are such that $xy \in f^{-1}(B')$, then $f(x)f(y) \in B'$, hence $x \notin f^{-1}(B')$ would imply $f(y) \in Q$ or $y \in P$; similarly, $y \notin f^{-1}(B')$ yields $x \in P$ and thus $\Omega \circ f$ defines a ϕ-pre-place of A/K. Consider

$\tilde{f} : \mathrm{Prim}_K(B) \to \mathrm{Prim}_K(A)$, given by $\tilde{f}(Q) = f^{-1}(Q)$. This map $\tilde{f}$ is continuous, because

$$\tilde{f}^{-1}(D(F)) = \{Q \in \mathrm{Prim}_K(B), \tilde{f}(Q) \in D(F)\} =$$
$$= \{Q \in \mathrm{Prim}_K(B),\ f^{-1}(Q) \cap F = \phi\}$$
$$= \{Q \in \mathrm{Prim}_K(B),\ Q \cap f(F) = \phi\} = D(f(F)).$$

Restricting $\tilde{f}$ to ϕ-$\mathrm{Prim}_K(A)$ yields maps $\tilde{f}_\phi$ which are continuous by definition of the ϕ-topology.

PROPOSITION 68. Every prime P of A/K contains a prime ideal P^0 of A; P^0 is the maximal element in the set of ideals of A contained in P.

PROOF. Put $P^0 = \{x \in A, A\,x\,A \subset P\}$, P^0 is an ideal of A. Suppose that there exist $a, b \notin P^0$ such that $a\,A\,b \subset P^0$. Then P contains neither $A\,a\,A$ nor $A\,b\,A$ and thus we may find an $a' \in A\,a\,A$ and a $b' \in A\,b\,A$ which are not in P. Let P be represented by $(A', \psi, A_1/K_1)$. Since $a'\,A\,b' \subset A\,a\,A\,b\,A \subset P \subset A'$, we have $a'b' \in A'$. Thus $a' \notin A'$ implies $b' \in P$ while $b' \notin A'$ implies $a' \in P$. Therefore both a' and b' are in A' but then $a'A'b' \subset P$ contradicts the fact that $A' - P$ is multiplicatively closed. Observe that P^0 is in general not completely prime. A prime of A/K which is also an ideal of A is a completely prime ideal of A. Obviously, for an arbitrary prime P of A/K, the set $A - P$ is multiplicatively closed.

The following characterization of symmetric primes generalizes a result of I. Connell [6].

PROPOSITION 69. Let ϕ be a place of K with valuation ring O_K and maximal ideal M_K. Then P is a proper symmetric ϕ-prime of A/K if and only if :

1. P is an O_K-module, such that $P \cap K = M_K$.
2. P is multiplicatively closed and symmetric, i.e., $Py \subset P$ is

equivalent to $yP \subset P$, for every $y \in A$.

3. The complement of P is multiplicatively closed.

PROOF. The only if part is obvious. Conversely let $A' = \{x \in A, xP \subset P\}$ then :

a) $A' \cap K = O_K$; since $\alpha \in K$ and $\alpha P \subset P$ entail $\alpha^{-1} \notin P$, hence $\alpha^{-1} \notin M_K$.

b) A' is an O_K-algebra containing P as an ideal.

c) Let $x,y \in A$ be such that $xy \in A'$ but $y \notin A'$.

Since P is symmetric, the fact that $xyP \subset P$ implies $Pxy \subset P$ while $yP \not\subset P$ and $Py \not\subset P$. Take $p \in P$ so that $yp \notin P$. Then $xyp \in P$ implies $x \in P$ because $A - P$ is multiplicatively closed. Similarly, $xy \in A'$ with $x \notin A'$ yields $y \in P$. Finally, since A' is an O_K-algebra and since $A - P$ is multiplicatively closed, it follows that P is completely prime in A' and thus P is a ϕ-prime of A/K.

From the remarks following Proposition 62 we derive :

PROPOSITION 70. If $(A',\psi,A_1/K_1)$ is a pre-place of A/K then $\tilde{\psi} : PR_{K_1}(A_1) \to PR_K(A)$ induces maps $\phi_1\text{-Prim}_{K_1}(A_1) \to \phi\text{-Prim}_K(A)$, where ϕ is the place of K defined as the specialization of $\psi|K$ obtained by composition with the place ϕ_1 of K_1. We also obtain a map, again denoted by $\tilde{\psi}$, $\text{Prim}_{K_1}(A_1) \to \text{Prim}_K(A)$. All mappings are continuous in the corresponding topologies.

The image $\text{Im}\,\tilde{\psi}$ of $\tilde{\psi} : \text{Prim}_{K_1}(A_1) \to \text{Prim}_K(A)$ is the set of <u>specializations</u> of the prime $P = \text{Ker}\,\psi$. If $P' \in \text{Im}\,\tilde{\psi}$ then we write $P \to P'$.

PROPOSITION 71. A necessary and sufficient condition for $P_1 \to P_2$ is that there exist representing pre-places ψ_1, ψ_2 for P_1, P_2 resp. with rings of definition A_1', A_2' resp., such that $A_2' \subset A_1'$ and $P_2 \supset P_1$. This is an immediate consequence of Proposition 62. A prime P of A/K represented by ψ is called a <u>minimal prime</u> if $P \in \text{Im}\,\tilde{\Omega}$ with $\Omega \in PR_K(A)$

implies $\Omega \cong \psi$.

Example. A minimal prime ideal which is completely prime is a minimal prime. More examples may be constructed over discrete rank one valuation rings, see Proposition 77.

Consider a K-algebra morphism $g : A \to B$. Let A_f be a subalgebra of A, containing K, such that if $xy \in A_f$ with $x,y \in A$ and $x \notin A_f$, then $y \in \text{Ker } f$ while $y \notin A_f$ yields $x \in \text{Ker } f$, then :

PROPOSITION 72. If $P \in \text{Prim}_K(A_f)$ and $P \supset \text{Ker } f$ then $P \in \text{Prim}_K(A)$.

PROOF. Let $(A_f', \psi, \bar{A}_f)$ represent P. If $x,y \in A$ and $xy \in A_f'$ but $y \notin A_f'$ then :

a) $y \notin A_f$; then $xy \in A_f$ implies that $x \in \text{Ker } f \subset P$.

b) $y \in A_f$; then $xy \in A_f'$ with $x \notin A_f$ yields $y \in \text{Ker } f$, contradiction. Hence let $x \in A_f$. But $x,y \in A_f$ and $xy \in A_f'$ with $y \notin A_f'$ entails $x \in P$ since P is a prime of A_f/K. Similarly, $xy \in A_f'$ with $x \notin A_f'$ yields $y \in P$, hence $P \in \text{Prim}_K(A)$.

In case f is surjective we have that primes P of A/K which contain Ker f yield primes f(P) of B/K and it is easily verified that $\tilde{f} : \text{Prim}_K(B) \to \text{Prim}_K(A)$ defines a homeomorphism of $\text{Prim}_K(B)$ onto $V(\text{Ker } f) = \{P \in \text{Prim}_K(A), P \supset \text{Ker } f\}$ which is Zariski-closed.

PROPOSITION 73. Let $\phi_1 \to \phi_2$ be places of K with associated valuation rings $O_1 \to O_2$ resp. Let P_i be a ϕ_i-prime of A/K represented by $(A_i', \psi_i, A_i/K_i)$ $i = 1,2$, and suppose $P_2 \to P_1$. Then $P_2 \cap A_1'$ is a ϕ_2-prime of A/K and a specialization of P_1.

PROOF. The situation may be summarized in :

$$P_1 \subset P_2 \cap A_1' \subset A_2' \cap A_1' \subset A_1' \subset A.$$

It is sufficient to show that $\psi_2|A_1'$ defines a pre-place on $A_2' \cap A_1'$ with kernel $P_2 \cap A_1'$.

If $x,y \in A$ are such that $xy \in A_1' \cap A_2'$ and $y \notin A_1' \cap A_2'$, then

a) If $y \notin A_1'$ then $xy \in A_1'$ yields $x \in P_1 \subset P_2 \cap A_1'$.

b) If $y \notin A_2'$ then $xy \in A_2'$ entails $x \in P_2$. If $x \notin A_1'$ then $xy \in A_1'$ entails $y \in P_1 \subset A_2'$ contradiction. Hence $x \in A_1' \cap P_2$.

In a similar way it follows that $xy \in A_1' \cap A_2'$ with $x \notin A_1' \cap A_2'$ implies $y \in A_1' \cap P_2$. Since $(A_1' \cap P_2) \cap K = M_2$ we are left to prove that $(A_1' \cap A_2') - (A_1' \cap P_2)$ is multiplicatively closed and for this it is enough to show that $A - (A_1' \cap P_2)$ is multiplicatively closed.

Let $x,y \in A$ be such that $xy \in A_1' \cap P_2$ and $y \notin A_1' \cap P_2$. Again, two possibilities occur :

a) $y \notin P_2$; then $xy \in P_2 \cap A_1'$ forces $x \in P_2$ because $A - P_2$ is multiplicatively closed. The assumption $x \notin A_1'$ with $xy \in A_1'$ would lead to $y \in P_1 \subset P$, hence $x \in A_1' \cap P_2$.

b) $y \notin A_1'$; then $xy \in A_1'$ entails $x \in P_1 \subset P_2 \cap A_1'$.

The fact that $P_2 \cap A_1'$ is specialization of P_1 is an easy corollary of Proposition 72.

In the above proof, the fact that P_1 is actually a prime is not used in full strength but merely that A_1' has the properties of A_f in Proposition 71.

With notations as in Proposition 71 we have the following

<u>COROLLARY</u>. If P is a ϕ-prime of A/K, $P \supset \text{Ker } f$, then $P \cap A_f$ is a ϕ-prime of A_f. Note that P is not necessarily a specialization of $P \cap A_f$.

Taking $\phi = \phi_1 = \phi_2$, $P = P_1 = P_2$ in Proposition 73, we see that there exists a pre-place of A/K representing P such that the ring where the pre-place is defined, is minimal with respect to inclusion. This particular representative for P will be referred to by $(A_p,\psi,A_1/K_1)$ and if this pre-place is special, restricted or unramified then P is said to be <u>absolutely</u> special, restricted or unramified.

PROPOSITION 74.

1. A prime P of A/K is special if and only if it is absolutely special.
2. A prime P of A/K is absolutely restricted if and only if every pre-place representing P is restricted.
3. If $[A : K] < \infty$, then a prime P of A/K is unramified if and only if P is absolutely unramified.
4. If A is a skew field, then P is a restricted prime of A/K if and only if P is absolutely restricted. Then there is a unique (up to isomorphism) $(A',\psi,A_1/K_1)$ representing P.

PROOF. 1. is trivial. Let P be restricted and let $(A_1',\psi_1,A_1/K_1)$ be a restricted representative for P. Let $(A_2',\psi_2,A_2/K_2)$ be any other representative for P. If $y \in A_2' - A_1'$ then $\lambda y \in A_1' - P$ with $\lambda \in M_K$. Then $y \in A_2'$ yields $\lambda y \in P$, contradiction, thus $A_2' \subset A_1'$ follows and this proves the assertion 2. To prove 4, let $(A',\psi,A_1/K_1)$ be a restricted pre-place of the skew field A representing P. In general $A_p \subset A'$. Suppose $y \in A' - A_p$. Then certainly $y \in A' - P$ and thus $y y^{-1} \in A'$ implies $y^{-1} \in A'$. Moreover $y^{-1} \in A' - P$ because if y^{-1} were in P then also $1 = y y^{-1} \in P$. However $y y^{-1} \in A_p$ with $y \notin A_p$ entails $y^{-1} \in P$, contradiction. Therefore A' and A_p coincide. Assertion 3 follows from 4, Proposition 65, 2, and the fact that a finite dimensional K-algebra which is a division ring is a skew field (see Lemma 78 corollary).

PROPOSITION 75. Let ψ be a restricted pre-place representing a prime P, then P is symmetric and $P = M_K \cdot (A' - P)$.

PROOF. Suppose that $yP \subset P$, $Py \not\subset P$ for some $y \in A$. Clearly $y \notin A'$ and thus there is a $\lambda \in M_K$ such that $\lambda y \in A' - P$. Since $\lambda y = y\lambda \in yP \subset P$ we reach a contradiction. Let $x \in P$. There is a $\lambda \in K$ such that $\lambda x \in A' - P$, thus $P^0 = (0)$. Moreover if $y = \lambda x \in A' - P$ then $\lambda \notin 0_K$ follows, hence $\lambda^{-1} \in M_K$ and $x = \lambda^{-1}y$.

Remark. From Proposition 58 follows that a restricted prime of A/K intersects any subfield L/K of A/K in a valuation ring of L. This shows that restricted primes generalize the concept of a valuation ring and that $\mathrm{Prim}_K(A)$ is in a way related to the Riemann surface of a field.

PROPOSITION 76. Let $[A : K] < \infty$, and let P_2 be an unramified ϕ_2-prime of A/K represented by $(A_2', \psi_2, A_2/K_2)$. Then for every place ϕ_1 of K such that $\phi_1 \to \phi_2$, there exists an unramified pre-place ψ_1 inducing ϕ_1 in K such that $P_2 \in \mathrm{Im}\,\tilde{\psi}_1$.

PROOF. The remarks following Proposition 60 yield that there exists an unramified pseudo-place $(A_1', \psi_1, A_1/K_1)$ such that $\psi_1 \to \psi_2$. Hence A_1 is a division ring. By Proposition 57, ψ_1 is a pre-place and since $\mathrm{Ker}\,\psi_1 \to P_2$ we have $P_2 \in \mathrm{Im}\,\tilde{\psi}$.

COROLLARY. To a specialization chain of places of K :

(*) $1 \to \phi_1 \to \ldots \to \phi_n = \phi$,

such that there exists an unramified ϕ-prime P of A/K, there corresponds a chain (**) of unramified primes of A/K :

(**) $P = P_n \supset P_{n-1} \supset \ldots \supset P_1 \supset 0$.

The chain (*) is maximal if and only if (**) is maximal.

Note that isomorphic unramified primes are equal. Thus, in case $[A : K] < \infty$, minimal unramified primes are necessarily defined over a Noetherian valuation ring. In the absence of the finiteness condition on $[A : K]$ we get,

PROPOSITION 77. Any absolutely restricted prime over a Noetherian valuation ring of K is minimal.

PROOF. Let P be absolutely restricted and let $P \cap K = 0_K$ be Noetherian, i.e., a discrete rank one valuation ring. Suppose that P_1 is a non-zero prime of A/K such that $P_1 \to P$, i.e. $0 \neq P_1 \subset P \subset A' \subset A_1'$, where A' and A_1' are subrings where P and P_1 resp. are defined. Since 0_K is a maximal subring of K we have that $A_1' \cap K = 0_K$ and thus $P_1 \cap K = M_K$. If $x \in P - P_1$ then $x \in A_1'$ and since P is restricted, there is a $\lambda \notin 0_K$ such that $\lambda x \in A' - P$. From $\lambda^{-1} \in M_K$ we derive that

$$x = \lambda^{-1} (\lambda x) \in M_K A_1' \subset P_1,$$

contradiction. Thus $P = P_1$. If ϕ is a Noetherian place of K then an absolutely restricted ϕ-prime of A/K is called a site of A/K.

LEMMA 78. Let $[A : K] < \infty$ and let $(A',\psi,A_1/K_1)$ be an unramified ϕ-pseudo-place of A/K such that A_1 is a prime ring. Then ψ is the unique ϕ-pseudo-place defined on A' such that the residue algebra is a prime ring.

PROOF. By Theorem 56, A' is a free 0_K-module of dimension $[A : K] = n$. A well-known fact for finite dimensional torsion free 0_K-algebras translates into : $\Sigma_\psi \dim_{K_1} \psi(A') \leq \dim_{0_K}(A') = n$, the sum ranging over ϕ-pseudo-places of A/K defined on A' with prime residue ring. Hence $(A',\psi,A_1/K_1)$ is unique as such.

COROLLARY. If A/K is a finite dimensional prime algebra then A is simple. Indeed $(A,1,A)$ is unramified, hence (0) is the unique prime ideal of A. If A is moreover a division ring, then it is a skew-field.

LEMMA 79. Let P be a restricted ϕ-prime of an arbitrary K-algebra A. If P is represented by $(A',\psi,A_1/K_1)$ such that A' is an 0_K-integral algebra then P is the unique restricted ϕ-prime of A/K.

PROOF. Suppose that $P_1 \subset A_1'$ is another restricted ϕ-prime, i.e., $A_1' \cap K = 0_K$, $P_1 \cap K = M_K$ where 0_K is the valuation ring of ϕ. First, let $x \in A' - A_1$. We have a relation $x^n + \ldots + a_{n-1}x + a_n = 0$ with $a_i \in 0_K$. Pick $\lambda \in M_K$ such that $\lambda x \in A_1' - P_1$, then $\lambda x \in P$. We obtain

$$(\lambda x)^n + \lambda a_1(\lambda x)^{n-1} + \ldots + \lambda^n a_n = 0,$$

where $\lambda a_1(\lambda x)^{n-1} + \ldots + \lambda^n a_n \in P_1$ because $\lambda x \in A_1'$ and $\lambda^i a_i \in M_K$. Hence $(\lambda x)^n \in P_1$, yielding $\lambda x \in P_1$, contradiction. Therefore $A' \subset A_1'$ and then $P \subset P_1$ by Proposition 75. Secondly, suppose $x \in A_1' - A'$. Pick $\lambda \in M_K$ such that $\lambda x \in A' - P$, then $\lambda x \in P_1$. Since $\lambda x \in A'$ we have a relation $(\lambda x)^n + \ldots + a_1(\lambda x) + a_0 = 0$ with $a_i \in 0_K$. Obviously

$$(\lambda x)^n + \ldots + a_1(\lambda x) \in P_1,$$

thus $a_0 \in P_1 \cap K = M_K$. Further,

$$(\lambda x)^n + \ldots + a_1(\lambda x) \in P$$

but since $\lambda x \in A' - P$ we get

$$(\lambda x)^{n-1} + \ldots + a_1 \in P \subset P_1$$

yielding $a_1 \in M_K$ and so on. Finally we arrive at $\lambda x + a_{n-1} \in P \subset P_1$ thus $a_{n-1} \in M_K$ and $\lambda x \in P$, contradiction. Therefore, $A' = A_1'$ and $P_1 = P$.

Let P be a site of a K-algebra A, represented by $(A_p,\psi,A_1/K_1)$. The maximal ideal $M_K = P \cap K$ of 0_K is principal, $M_K = (m)$ say. We define a function $\mathrm{ord}_p : A \to \mathbb{Z}$, by $\mathrm{ord}_p(x) = -n$ if and only if $m^n x \in A_p - P$.

PROPOSITION 80. If P is a site of A/K then ord_p is an order function on A, i.e. ord_p satisfies :

1. $\mathrm{ord}_P(x + y) \geq \min(\mathrm{ord}_P(x), \mathrm{ord}_P(y))$
2. $\mathrm{ord}_P(x\,y) = \mathrm{ord}_P(x) + \mathrm{ord}_P(y)$.

PROOF. Note that $P = \{x \in A, \mathrm{ord}_P(x) > 0\}$ and $A_P = \{x \in A, \mathrm{ord}_P(x) \geq 0\}$.

1. Put $\mathrm{ord}_P(x + y) = -n$. Then $m^n(x + y) = m^n x + m^n y \in A_P - P$ yields that at least one of the elements $m^n x$, $m^n y$ is not in P, so $-n \geq \mathrm{ord}_P(x)$ say. If $-n = \mathrm{ord}_P(x)$ then $m^n y \in A_P$ yields $-n \leq \mathrm{ord}_P(y)$. If $-n > \mathrm{ord}_P(x)$ then $m^n y \notin A_P$ or, $-n > \mathrm{ord}_P(y)$. Thus $\mathrm{ord}_P(x + y) \geq \min(\mathrm{ord}_P(x), \mathrm{ord}_P(y))$.

2. Let $\mathrm{ord}_P(x\,y) = -n$, i.e., $m^n xy \in A_P - P$. Put $\mathrm{ord}_P(x) = -N$. Then $m^{n-N} y\, m^N x \in A_P - P$ and $m^{n-N} y \notin P$ because $m^N x \in A_P$. Moreover $m^{n-N} y \in A_P$ because $m^N x \notin P$ and P being a prime. Thus $\mathrm{ord}_P(y) = N - n$ and $-n = (N - n) + (-N)$ or $\mathrm{ord}_P(x\,y) = \mathrm{ord}_P(x) + \mathrm{ord}_P(y)$.

Remark. The assumption that P is absolutely restricted may be keyed down to P being restricted. However, since in that case the choice of the ring A' defining the prime is not prescribed, we obtain order functions $\mathrm{ord}_{(P,A')}$. In case A is a finite dimensional K-algebra, sites are related to (maximal) orders in skew fields. From now on in this section, let $[A : K] = n < \infty$. Sites of A/K can only exist when A is a division ring, thus only when A is a skew field, by Lemma 78, Corollary.

PROPOSITION 81. If $(A', \psi, A_1/K_1)$ represents a site P of a skew field A/K, then A' is a valuation ring of A.

PROOF. Suppose $x, x^{-1} \in A$ are such that $x, x^{-1} \notin A'$. Since a site is restricted, $\lambda x \in A' - P$ and $\mu x^{-1} \in A' - P$ for some $\lambda, \mu \in M_K$. But $\lambda\mu \in A' - P$ contradicts $\lambda\mu \in M_K$, hence if $x \notin A'$ then $x^{-1} \in A'$.

COROLLARY. If K is complete with respect to the valuation corresponding to the valuation ring O_K then, if (A', ψ, A_1, K_1) represents a site of

A/K, A' is an integral 0_K-algebra and thus a maximal order of A/K.

PROPOSITION 82. Let $(A',\psi,A_1/K_1)$ be a representative of a minimal unramified prime P of A/K, then A' is a maximal order and a maximal subring of A.

PROOF. Since minimal unramified primes are necessarily defined over a Noetherian valuation ring of K, we have that $A' \cap K = 0_K$ is Noetherian. Theorem 56 yields that $A' = 0_K[e_1,\dots,e_n]$ for some K-basis $E = \{e_1,\dots,e_n\}$ of A with $e_i \in A' - P$ for all i. Any $x \in A'$ gives rise to an ascending chain of 0_K-submodules in A' : $x\,0_K \subset (x\,0_K, x^2 0_K) \subset \dots$, which is stationary. This yields an integral relation for x over 0_K, thus A' is an integral 0_K-algebra, and thus an order of A. Lemma 79 entails that P is the unique restricted ϕ-prime of A/K, where ϕ denotes the place of K associated to 0_K. Over a Noetherian valuation ring, every order is contained in a maximal order, so let M be a maximal order in A/K and $A' \subset M$. Suppose there exists an element $y \in M - A'$. Since $y = \sum_{i=1}^{n} y_i e_i$, with $y_i \in K$ we get that $y_j^{-1} y = y^*$ is in A' for some j, $1 \leq j \leq n$. By Proposition 81, $y^* \notin P$ implies that $(y^*)^{-1}$ is in A', hence $y_j = y(y^*)^{-1} \in M \cap K$. However, $M \cap K$ cannot contain 0_K properly unless $K \subset M$, proving $A' = M$.

IV. 5. Localization at Primes, and Sheaves.

Let P be a prime of A/K which is represented by the pre-place $(A',\psi,A_1/K_1)$. Consider the following symmetric kernel functors :

1. $\sigma_{A'-P}$ is the symmetric kernel functor on M(A) correspoding to the multiplicative set A' - P.
2. σ_{A-P} is the symmetric kernel functor on M(A) corresponding to the multiplicative set A - P.
3. $\sigma_{A'}$ is the symmetric kernel functor on M(A') associated to the multiplicative set A' - P.

<u>LEMMA</u> 82. $C(\sigma_{A-P}) = \{P^0\}$, hence $\sigma_{A-P} = \sigma_{A-P^0}$ where σ_{A-P^0} is the symmetric kernel functor associated to the prime ideal P^0 of A (as in I. 1.).

<u>PROOF</u>. P^0 obviously is the unique ideal, maximal with the property of being disjoint from $A-P$, hence $C(\sigma_{A-P}) = \{P^0\}$ and because σ_{A-P} is symmetric this implies $\sigma_{A-P} = \sigma_{A-P^0}$.

Via $A' \hookrightarrow A$ we identify $M(A)$ with a subcategory of $M(A')$. However $\sigma_{A'}$, when acting on A-modules does not necessarily define a kernel functor on $M(A)$. If ψ is special, then it is easily checked that $\sigma_{A'}$ induces an idempotent kernel functor on $M(A)$, moreover,

<u>LEMMA</u> 83. If ψ represents a special prime P of A/K, then $\sigma_{A'}$ and $\sigma_{A'-P}$ coincide on $M(A)$.

<u>PROOF</u>. Take $M \in M(A)$. By definition of the A'-module structure on M, $\sigma_{A'-P}(M) \subset \sigma_{A'}(M)$. If, $x \in \sigma_{A'}(M)$, i.e., $sA'x = 0$ for some $s \in A'-P$ then look at sAx. Suppose there is a non-zero $y \in sAx$. We may find $\lambda \neq 0$ in K such that $\lambda y = \lambda s a x = s(\lambda a)x$ is in $sA'x = 0$, hence $y = 0$, contradiction. Thus x is also an element of $\sigma_{A'-P}(M)$.

<u>PROPOSITION</u> 84. Let ψ represent a special prime P of A/K. Then A' is a prime ring if and only if A is a prime ring; A is $\sigma_{A'}$-torsion free if and only if A' is $\sigma_{A'}$-torsion free.

The proof is straightforward.

<u>PROPOSITION</u> 85. If ψ is a restricted pre-place representing the prime P, then $\sigma_{A'}$, $\sigma_{A'-P}$ and σ_{A-P} coincide with σ^* on $M(A)$.

<u>PROOF</u>. Since P is special, $\sigma_{A'}$ and $\sigma_{A'-P}$ coincide on $M(A)$. Since P is restricted $P^0 = (0)$, hence $C(\sigma_{A-P}) = \{(0)\}$ or $\sigma_{A-P} = \sigma^*$ on $M(A)$.

We are left to prove that $\sigma_{A-P} = \sigma_{A'-P}$ on $M(A)$. Let $I \in T(\sigma_{A-P})$, i.e., $I \supset (s)$ with $s \in A - P$. There may be found $\lambda \in K$ such that $\lambda s \in A' - P$, but then from $(\lambda s) \subset I$ it follows that I is also in $T(\sigma_{A'-P})$. The inequality $\sigma_{A'-P} \leqslant \sigma_{A-P}$ follows directly from $A' - P \subset A - P$.

Remark. Denote $\sigma_{A'}$ by σ. If P is restricted, then A, hence also A', are prime rings. Moreover A and A' are σ^*- and σ-torsion free respectively. In general, if P is any prime of A/K for which we have injective maps $A' \hookrightarrow Q_\sigma(A') \hookrightarrow Q_\sigma(A)$, then $A \cap Q_\sigma(A') = A'$. Indeed, if $y \in Q_\sigma(A') - A'$ then for some ideal $L \in T(\sigma)$, $L \supset (s)$ with $s \in A' - P$ we have $Ly \subset A'$. But $sy \in A'$ with $s \in A' - P$ yields $y \notin A$. In this situation if $Q_\sigma(A) = A$ then $Q_\sigma(A') = A'$. If ψ is restricted and $[A : K] > \infty$ then $Q_\sigma(A) = A$, $Q_\sigma(A') = A'$ and A is a skew field.

Put $X = \mathrm{Prim}_K A$ and let E be a finite subset of A, X_E the corresponding basic open set. To X_E we associate $\overline{E} = \cap \{A - P, P \in X_E\}$, which is a multiplicative subset of A containing E. To the system $\overline{E}$ we associate a symmetric kernel functor σ_E. In this way we obtain a presheaf on X. Indeed, if $X_E \subset X_F$ then $\overline{F} \subset \overline{E}$ and thus $\sigma_F \leqslant \sigma_E$, a ring homomorphism $\rho(F,E) : Q_F(A) \to Q_E(A)$ results, where Q_F, Q_E are the localization functors with respect to σ_F, σ_E resp. (see Theorem 10). Furthermore, if $X_E \subset X_F \subset X_G$, then we obtain a commutative diagram of ring homomorphisms :

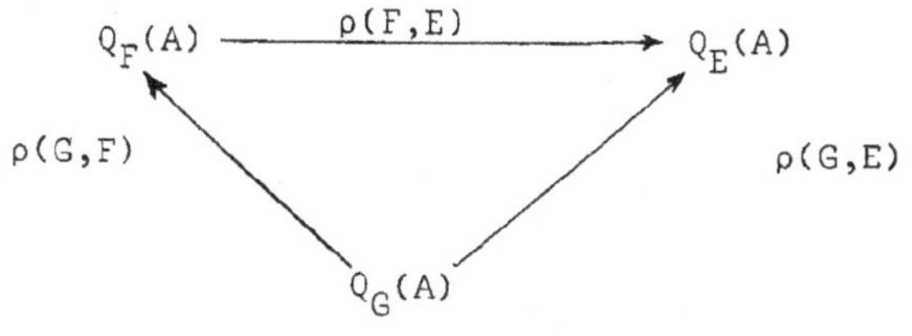

where $\rho(F,E)$ is the unique extension of the canonical $A \to Q_E(A)$ to a ring homomorphism $Q_F(A) \to Q_E(A)$; similar for $\rho(G,F)$, $\rho(G,E)$. Sheafification of the presheaf yields a sheaf $\tilde{Q}$ on X.

<u>THEOREM</u> 86. The stalk of $\tilde{Q}$ at $P \in X$ is exactly $Q_{A-P}(A)$.

<u>PROOF</u>. The direct limit of the system $\{Q_U(A), \rho(U,V), V \subset U\}$ equals the direct limit of the cofinal system $\{Q_E(A), \rho(E,F), X_F \subset X_E\}$. Since the functors σ_E are symmetric and associated to the multiplicative systems $\bar{E}$, the direct limit localization functor corresponds to the sup of the symmetric kernel functors involved. The latter kernel functor corresponds to the multiplicative system generated by $\cup\{\bar{E}, P \in X_E\}$, which is exactly $A - P$, becuase if $s \in A - P$ then $P \in X_s$ and

$$s \in \overline{\{s\}} \subset \cup\{\bar{E}, P \in X_E\}.$$

Let $f : A_1 \to A_2$ be a K-algebra morphism. Then $\tilde{f}^{-1}(X_E) = X_{f(E)}$. If $s \in f(\bar{E})$, let $x \in \bar{E}$ be so that $f(x) = s$, then for $P' \in X_{f(E)}$ we have that $f^{-1}(P') = \tilde{f}(P') \in X_E$ and $x \notin f^{-1}(P')$. This entails $s \notin P'$ and $f(\bar{E}) \subset \overline{f(E)}$. If f is surjective then the foregoing yields, by σ_E-injectivity, a map $Q_E(A_2) \to Q_{f(E)}(A_2)$ which composes with $Q_E f : Q_E(A_1) \to Q_E(A_2)$ to give a morphism $Q_E(A_1) \to Q_{f(E)}(A_2)$. To get that these morphisms are ring homomorphisms, compatible with sheaf restriction maps, more conditions have to be imposed on the K-algebra A (for example : hereditary left Noetherian rings with Artin-Rees condition). In the commutative case, Prim_K is a functor $\underline{\text{Alg}}_K \to \underline{\text{Sheaves}}$, generalizing Spec and Gam_K. This follows from the fact that $\text{Prim}_K(A) \supset \text{Spec } A$ and, if $E = \{e_1,\dots,e_n\} \subset A$ and $\langle e\rangle$ being the multiplicative set generated by $e = e_1 \dots e_n$, then

$$\langle e\rangle \subset \bar{E} \subset A - \cup\{\text{Spec } A, e \notin P\}.$$

Hence, the right hand set being the saturation of $\langle e\rangle$, it follows that $\langle e\rangle^{-1} A \cong Q_E(A)$. The rest of the proof is verification of the compatibility with sheaf restrictions, along the lines of [6], Proposition 24.

Special references for Section IV.

M. AUSLANDER, O. GOLDMAN [2], G. AZUMAYA [3], I.G. CONNELL [6], M. DEURING [7], I.N. HERSTEIN [14], F. VAN OYSTAEYEN [34], [35], [37].

V. APPLICATION : THE SYMMETRIC PART OF THE BRAUER GROUP

V. 1. Generic Central Simple Algebras.

For finite abelian groups G we aim to construct "functors" $\mathcal{D}(G)$ from the category of fields with the morphisms being the surjective places, to G-crossed product skew fields with galoisian pseudo-places for the morphisms.

The skew fields $\mathcal{D}_K(G)$ thus obtained have a generic property as defined on page 66. Let $\psi : \mathcal{D}_K(G) \to \mathcal{D}_k(G)$ be the galoisian pseudo-place corresponding to a place $\phi : K \to k$, then Im ψ contains the set S of all specializations of ψ obtained by composition with galoisian pseudo-places of $\mathcal{D}_k(G)$ such that every crossed product $(G,1/k,\{C_{\sigma,\tau}\})$ defined by a symmetric factor set $\{C_{\sigma,\tau}\}$ is residue algebra under some element of S. Taking $\phi = \mathrm{Id}_K$, it follows that $\mathcal{D}_K(G)$ has to be constructed in such a way that every crossed product $(G,1/k,\{C_{\sigma,\tau}\})$ is residue algebra of $\mathcal{D}_K(G)$ under a galoisian pseudo-place. In this section, "generic" stands for the above described $H^2_{sym}(G,K^*)$-generic property.

LEMMA 87. Let G be a finite abelian group of order n and exponent m. Let k be a field the characteristic of which does not divide n and containing the m-th roots of unity. Let G act trivially in $K = k(t_1,\dots,t_n)$, a purely transcendental extension of k of transcendence degree n. Then there exist fixed elements $(\sigma_1,\tau_1),\dots,(\sigma_n,\tau_n)$ in $G \times G$ and a symmetric cocycle f of G in $K(t_1,\dots,t_n)^*$ such that $f(\sigma_i,\tau_i) = t_i$, $i = 1,\dots,n$, and $f(\sigma,\tau)$ is in the free multiplicative subgroup S of $K(t_1,\dots,t_n)^*$ generated by $\{t_1,\dots,t_n\}$.

PROOF. The lemma appears in a slightly different form (and context) in [16].

Since this provides a universal formula for the expression of $f(\sigma,\tau)$ in terms of the pre-selected $f(\sigma_i,\tau_i)$, it follows that one easily generalizes the previous lemma to the case where k is arbitrary. Any solution $\{C_{\sigma,\tau}, \sigma,\tau \in G\}$ of the cocycle relations :

$$(*)\quad \begin{cases} x_{\sigma,\tau}\, x_{\sigma\tau,\rho} = x_{\sigma,\tau\rho}\, x_{\tau,\rho} \\ x_{\rho,\tau} = x_{\tau,\rho} \end{cases}$$

in k^*, yields a map $t_i \to C_{\sigma_i,\tau_i}$ which extends to a homomorphism $h_c : S \to k^*$ with $C_{\sigma,\tau} \in h_c(S)$ for all $\sigma,\tau \in G$. From h_c we derive $h_c^* : H^2(G,S) \to H^2(G,k^*)$ with $h_c^*(f) = C$, $C = [\{C_{\sigma,\tau}\}]$, $f = [\{x_{\sigma,\tau}\}]$ where $x_{\sigma,\tau}$ denotes a solution of (*) with $x_{\sigma_i,\tau_i} = t_i$. Since Hom$(S,k^*)$ is in one-to-one correspondence with set-mappings $\{t_1,\ldots,t_n\} \to k^*$, we obtain a method for finding all symmetric factor sets $\{C_{\sigma,\tau}\}$ definining a class $C \in H^2(G,l^*)$ where l/k is any galoisian field extension with Gal$(l/k) \cong G$, as follows. Every specialization $t_i \to \alpha_i \in k^*$, yields a symmetric cocycle defined by $C_{\sigma_i,\tau_i} = \alpha_i$, while vice-versa, every factor set representing a class $C \in H^2\mathrm{sym}(G,l^*)$ is obtained in this way. The method may be ameliorated in the following way.

<u>PROPOSITION</u> 88. Let G be an abelian group of order n, $G \cong C_1\times\ldots\times C_r$, C_i being cyclic of order n_i, $i = 1,\ldots,r$. Let k be any field and let S be the free abelian group generated by $\{t_1,\ldots,t_r\}$. Then there is a fixed $f \in H^2(G,S)$ such that for every $\alpha \in H^2(G,k^*)$ a group homomorphism $h_d : S \to k^*$ exists, such that, under the derived homomorphism $h_d^* : H^2(G,S) \to h^2(G,k^*)$ the fixed element f is mapped onto d.

<u>PROOF</u>. Let $\{C_{\sigma,\tau}\}$ represent d and let C_j be generated by σ_j. Put

$$d_j = \prod_{e=1}^{n_j} C_{\sigma_j,\sigma_j^e} \qquad j = 1,\ldots,r$$

and $d_{\sigma,\tau} = d_1^{\varepsilon_1}\ldots d_r^{\varepsilon_r}$ if $\sigma = \sigma_1^{v_1}\ldots\sigma_r^{v_r}$, $\tau = \sigma_1^{w_1}\ldots\sigma_r^{w_r}$, $\varepsilon_i = 1$ if $v_i + w_i \geq n_i$ and $\varepsilon_i = 0$ otherwise. The factor sets $\{C_{\sigma,\tau}\}$ and

$\{d_{\sigma,\tau}\}$ are equivalent. Define $f_{\sigma,\tau} = t_1^{\varepsilon_1} \cdots t_r^{\varepsilon_r}$ with $\sigma, \tau, \varepsilon_i$ as before, then the map $t_j \to d_j$, $j = 1, \ldots, r$, extends to a homomorphism $h_d : S \to k^*$ and h_d^* has the property $h_d^*(f) = d$.

Let G be a finite abelian group; l/k an arbitrary galoisian extension with $\mathrm{Gal}(l/k) \cong G$. Suppose that O_l is the valuation ring of an unramified place of l, then, since O_l and its maximal ideal M_l are globally G-invariant, we get an injective homomorphism :

$$H^2(G, O_l^*) \to H^2(G, l^*).$$

The inclusion $k^* \subset l^*$ yields a morhphism ϕ_l, $\phi_l : H^2(G,k^*) \to H^2(G,l^*)$, and a commutative diagram

$$\begin{array}{ccc} H^2(G,k^*) & \longrightarrow & H^2(G,l^*) \\ \uparrow & & \uparrow \\ H^2(G,k \cap O_l^*) & \longrightarrow & H^2(G,O_l^*) \end{array}$$

Now, the existence of such "generic" cocycles as obtained in Proposition 88 translates into the existence of generic crossed product algebras. So we change from the level of cohomology to the level of simple algebras.

Let $\mathrm{Gal}(l/k) = G$ be a finite abelian group of order n and add variables $\{t_1, \ldots, t_n\} = t$ to l, G acting trivially on t. The equation (*) with values ranging in $k(t)^*$ may be solved in the multiplicative group T generated by t, such that $x_{\sigma_i,\tau_i} = t_i$, the pairs (σ_i, τ_i) being preselected as in Lemma 87. This solution, aigain written $\{x_{\sigma,\tau}\}$ defines a crossed product $\mathcal{D}_{l/k}(G)$ of G with $l(t)$. It may be looked upon as being the algebra $l(t)[U_\sigma, \sigma \in G]$ with $U_\sigma U_\tau = x_{\sigma,\tau} U_{\sigma\tau}$, $U_\sigma U_\tau = U_\tau U_\sigma$ and $U_\sigma \lambda = \lambda^\sigma U_\sigma$ for all $\lambda \in l(t)$. For every crossed product $A = (G, l/k, \{C_{\sigma,\tau}\})$ with symmetric $\{C_{\sigma,\tau}\}$, there exists a galoisian pseudo-place of $\mathcal{D}_{l/k}$ with residue algebra A.

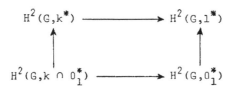

This is an easy consequence of the fact that we may write $A = l[u_\sigma, \sigma \in G]$, the $\{u_\sigma, \sigma \in G\}$ satisfying relations similar to the relations for $\{U_\sigma, \sigma \in G\}$, thus, specializing $x_{\sigma,\tau}$ to $C_{\sigma,\tau}$ and U_σ to u_σ we get what we want.

To an abelian group G of order n and a field k we associate the field $L = k(\{X_\sigma, \sigma \in G\})$ where $X = \{X_\sigma, \sigma \in G\}$ is a set of variables such that $\tau X_\sigma = X_{\tau\sigma}$, $\sigma,\tau \in G$, represents the action of G on X. Denote the fixed field for action of G in L by k_G. Clearly, $Gal(L/k_G) \cong G$ and every galoisian field extension l/k with $Gal(l/k) \cong G$ may be found as the residue field of L under an unramified place of L. This easily yields that every crossed product $A = (G, l/k, \{C_{\sigma,\tau}\})$ with symmetric $\{C_{\sigma,\tau}\}$ and arbitrary extension l/k such that $Gal(l/k) \cong G$, is residue algebra of $\mathcal{D}_k(G)/k_G(t)$ ($\mathcal{D}_k(G)$ stands for $\mathcal{D}_{L/k_G(t)}(G)$). Now, if $\phi : K \to k$ is a place of K, then a galoisian pseudo-place $\psi(K,k)$, $\psi(K,k) : \mathcal{D}_K(G) \to \mathcal{D}_k(G)$, may easily be derived from ϕ (replacing K by k in the construction of $\mathcal{D}_K(G)$). A crossed product $A = (G, l/k, \{C_{\sigma,\tau}\})$ is residue algebra of $\mathcal{D}_k(G)$ and also of $\mathcal{D}_K(G)$ under galoisian pseudo-places ψ_k and ψ_K respectively. We have then that

$$\tilde{\psi}(K,k) : PS_{k_G}(\mathcal{D}_k(G)) \to PS_{K_G}(\mathcal{D}_K(G))$$

maps ψ_k onto ψ_K, moreover, every galoisian pseudo-place Ω_K of $\mathcal{D}_K(G)$ with as residue algebra some crossed product $A = (G, l/k, \{C_{\sigma,\tau}\})$ is obtained in this way. For more properties of $\mathcal{D}_K(G)$ cf. [35], [17].

Proposition 88 allows us to construct another generic crossed product, using less variables in the process.

Again, $G = Gal(l/k)$ is abelian of order n. Let $G = C_1 \times \ldots \times C_r$ be a cyclic decomposition of G. Add r K-algebraic independent variables $\{t_1', \ldots, t_r'\} = t'$ to l and let G act trivially on t'.

Consider $\mathcal{S}_{l/k}(G) = l(t')[V_1, \ldots, V_r]$ with $V_i^{n_i} = t_i'$, n_i being the order of C_i.

The elements $\prod_{i=1}^{r} V_i^{e_i}$, $0 \leq e_i \leq n_i$ are $l(t')$-independent and V_i acts on $l(t')$ as it should, $V_i\lambda = \lambda^{\sigma_i} V_i$ for all $\lambda \in l(t')$, where σ_i is a generator of C_i. It is clear that $S_{l/k}(G)$ is a crossed product such that its factor set is obtained from $t_1',\ldots,t_r'$ in the same way $d_{\sigma,\tau}$ was expressed in terms of $\{d_1,\ldots,d_r\}$ in Proposition 88. Also, in case $L = k(\{X_\sigma, \sigma \in G\})$ we write $S_k(G)$ for $S_{L/k_G}(G)$. The properties of the generic algebras $\mathcal{D}_k(G)$, stated before fold if $\mathcal{D}$ is replaced by S everywhere, proofs have to be modified in that well-fitting factor sets are chosen so that Proposition 88 may be applied. There exists a galoisian pseudo-place of $\mathcal{D}_{l/k}(G)$ with $S_{l/k}(G)$ as residue algebra, cf. [35], hence from Proposition 65 it follows that $\mathcal{D}_{l/k}(G)$ is a skew field if $S_{l/k}(G)$ is a skew field.

PROPOSITION 89. The exponent of $S_{l/k}(G)$ in $\mathcal{B}r(k(t'))$ equals the exponent of $\mathcal{D}_{l/k}(G)$ in $\mathcal{B}r(k(t))$.

PROOF. Put $K = k(t_1,\ldots,t_n)$ and let $K(u_1,\ldots,u_r) = K(u)$ be the center of $S_{L/K}(G)$ where $L = l(t,u)$. The generic property of $S_{L/K}(G)$ yields a galoisian pseudo-place $(S_{L/K}(G), \psi_u, \mathcal{D}_{l/k}(G)/K)$. Since there is a galoisian pseudo-place, (cf. [35]) $(\mathcal{D}_{l/k}(G), \psi_t, S_{l/k}(G)/k(t'))$ we derive from Proposition 66 that $e(\mathcal{D}_{l/k}(G))$ divides $e(S_{L/K}(G))$ and that $e(S_{l/k}(G))$ divides $e(\mathcal{D}_{l/k}(G))$, e denoting the exponent in the corresponding Brauer groups. Since $S_{L/K}(G) \cong S_{l/k}(G) \otimes_{k(t')} k(u,t)$, this isomorphism derives from $k(u) \cong k(t')$ while $k(u,t)/k(u)$ is purely transcendent, we have that $e(S_{l/k}(G)) = e(S_{L/K}(G))$.

V. 2. Two theorems on generic crossed products.

We maintain notations introduced in V. 1..

THEOREM 90. The crossed product algebra $S_{l/k}(G) = l(t)[V_\sigma, \sigma \in G]$ is a skew field with center $k(t)$.

PROOF. The proof is based on the following. Let S be any division ring, σ an automorphism of S. Consider the noncommutative polynomial ring $S[T_\sigma]$ with multiplication rule $T_\sigma x = x^\sigma T_\sigma$, all $x \in S$. It is known that $S[T_\sigma]$ is a left (and right) principal ideal domain because a left (and right) algorithm exists. If the principal left ideal (f) of $S[T_\sigma]$ is an ideal then $S[T_\sigma]$ mod (f) is a division ring if and only if f is irreducible in $S[T_\sigma]$. Let $G = C_1 \times \ldots \times C_r$, C_i cyclic of order n_i generated by σ_i, and put $H_i = C_1 \times \ldots \times C_i$, $i = 1,\ldots,r$. We obtain a chain of field extensions : $k = l_r \subset l_{r-1} \subset \ldots \subset l_1 \subset l = l_0$, where l_i is the subfield of l left fixed by H_i. Hence $Gal(l/l_i) = H_i$, $Gal(l_i/k) \cong C_{i+1} \times \ldots \times C_r$, $i = 0,\ldots,r$. We are going to construct division algebras S_i , $i = 1,\ldots,r$, with center $l_i(t_1,\ldots,t_i)$ such that $S_0 = l \subset S_1 \subset \ldots \subset S_{r-1} \subset S_r = \mathcal{S}_{l/k}(G)$, and the construction of the S_i is such that, if S_i is a division algebra then $S_i + 1$ is too, $i \leq r - 1$. Put $t = \{t_1,\ldots,t_r\}$; add $t_1 \in t$ to k and look at $l(t_1)[T_{\sigma_1}]$ with action defined by $\lambda^{\sigma_1} T_{\sigma_1} = T_{\sigma_1} \lambda$ for all $\lambda \in l(t_1)$. Hence elements of $k(t_1)$ commute with T_{σ_1} and since $T_{\sigma_1}^{n_1} - t_1$ is in the center of $l(t_1)[T_{\sigma_1}]$ the left ideal P_1 generated by $T_{\sigma_1}^{n_1} - t_1$ is an ideal. Put $S_1 = l(t_1)[T_{\sigma_1}]/P_1$, thus $S_1 = l(t_1)[V_{\sigma_1}]$ where V_{σ_1} is the P_1-residue of T_{σ_1}. Repeating the process starting from S_1, and so on, we obtain $l = S_0, S_1, \ldots, S_{i-1}$, $i < r$. The algebra S_i is then constructed as follows. Extending the center of S_{i-1} we get $S'_{i-1} = S_{i-1} \otimes_{l_{i-1}(t_1,\ldots,t_{i-1})} l_{i-1}(t_1,\ldots,t_i)$.

We construct the skew polynomial ring $S'_{i-1}[T_{\sigma_1}]$ such that

$T_{\sigma_i} V_{\sigma_j} = V_{\sigma_j} T_{\sigma_i}$ for $i \leq j \leq i-1$, and $T_{\sigma_i} \lambda = \lambda^{\sigma_i} T_{\sigma_i}$ for all $\lambda \in l(t_1,\ldots,t_i)$. The left ideal P_i of $S'_{i-1}[T_{\sigma_i}]$ generated by $T_{\sigma_i}^{n_i} - t_i$ is an ideal; we put $S_i = S'_{i-1}[T_{\sigma_i}]/P_i$. Thus $S_i = l(t_1,\ldots,t_i)[V_{\sigma_1},\ldots,V_{\sigma_i}]$. Obviously $S_r \cong \mathcal{S}_{l/k}(G)$. <u>Now</u>, <u>if S_{i-1}, $1 \leq i < r$, is a division algebra then S_i is a division algebra</u>. Since

S_{i-1} is $l_{i-1}(t_1,\dots,t_{i-1})$-central and because it contains the galoisian extension $l(t_1,\dots,t_{i-1})$ of $l_{i-1}(t_1,\dots,t_{i-1})$, it follows that S_{i-1} is a crossed product, $S_{i-1} = (H_{i-1}, l(t_1,\dots,t_{i-1})/l_{i-1}(t_1,\dots,t_{i-1}), \{C_{\sigma;\tau}\})$ where $\{C_{\sigma,\tau}\}$ is the factor set defined by the $l(t_1,\dots,t_{i-1})$-basis $\{\prod_i V_{\sigma_i}^{e_i},\ 0 \leq e_i < n_i\}$ for S_{i-1}, hence $\{C_{\sigma,\tau}\}$ is symmetric and we have $V_{\sigma_j}^{n_j} = t_j$, $j \leq i-1$ while $V_{\sigma_p}V_{\sigma_q} = 1.V_{\sigma_p\sigma_q}$. This factor set will be referred to by $\{t\}_{i-1}$. Then S'_{i-1} is a division algebra and a crossed product $(H_{i-1}, l(t_1,\dots,t_i)/l_{i-1}(t_1,\dots,t_i), \{t\}_{i-1}\}$, if $i = 1$ then S'_0 is simply $l(t_1)$. It is sufficient to prove that $f = T_{\sigma_i}^{n_i} - t_i$ is irreducible in $S'_{i-1}[T_{\sigma_i}]$ because then S_i is a division algebra. Moreover, since V_{σ_i} commutes with $l_i(t_1,\dots,t_i)$, $Gal(l/l_i) = H_i$, we have that S_i contains $l(t_1,\dots,t_i)/l_i(t_1,\dots,t_i)$ as a maximal commutative subfield, so S_i is a crossed product $(H_i, l(t_1,\dots,t_i)/l_i(t_1,\dots,t_i), \{t\}_i)$. Assume $g,h \in S'_{i-1}[T_{\sigma_i}]$, both non-trivial and such that $f = gh$. The coefficients of g,h are rational functions of t_i with coefficients ranging over the division algebra S_{i-1}.

Choose $a \in S'_{i-1}$ such that :

$$a\,f = \left(\sum_{j=0}^{n_i} a_j(t_i)T_{\sigma_i}^j\right)\left(\sum_{s=0}^{n_i-1} b_s(t_i)T_{\sigma_i}^s\right),$$

where $a_j(t_i)$, $b_s(t_i) \in S_{i-1}[t_i]$ (this can be done with $a \in l_{i-1}(t_1,\dots,t_i)$). Modulo $(T_{\sigma_i}^{n_i} - t_i)$ this yields

$$(*)\qquad 0 = \left(\sum_{j=0}^{n_i-1} a_j(t_i)V_{\sigma_i}^j\right)\left(\sum_{s=0}^{n_i-1} b_s(t_i)V_{\sigma_i}^s\right).$$

Replacing t_i in the coefficients by $V_{\sigma_i}^{n_i}$ yields that the right-hand side of (*) is the product $g(V_{\sigma_i})h'(V_{\sigma_i})$ of non-trivial polynomials in $S_{i-1}[V_{\sigma_i}]$. If we show that $S_{i-1}[V_{\sigma_i}] \cong S_{i-1}[T_{\sigma_i}]$ then (*) is a contradiction. Therefore we show that the ideal $(f) \cap S_{i-1}[V_{\sigma_i}]$ of $S_{i-1}[T_{\sigma_i}]$ is the zero ideal, i.e. $S'_{i-1}[T_{\sigma_i}] \to S'_{i-1}[V_{\sigma_i}]$ restricts to an isomorphism $S_{i-1}[T_{\sigma_i}] \to S_{i-1}[V_{\sigma_i}]$. This is equivalent to showing that no relation of the form :

$$(**) \qquad g(t_i,T_{\sigma_i})(T_{\sigma_i}^{n_i} - t_i) = h(T_{\sigma_i}), \quad \text{with}$$

$$g(t_i,T_{\sigma_i}) \in S'_{i-1}[T_{\sigma_i}],\ 0 \neq h(T_{\sigma_i}) \in S_{i-1}[T_{\sigma_i}],$$

exists. The above relation reduces, up to multiplication by a suitable polynomial $C(t_i) \in S_{i-1}[t_i]$, to

$$(*^**) \qquad g^*(t_i,T_{\sigma_i})(T_{\sigma_i}^{n_i} - t_i) = C(t_i)h(T_{\sigma_i}), \text{ with}$$

$$g(t_i,T_{\sigma_i}) \in S_{i-1}[t_i,T_{\sigma_i}].$$

We write, $g^*(t_i,T_{\sigma_i}) = \sum_{j=0}^{m} a_j^*(t_i)\, T_{\sigma_i}^j$, with $a_j^*(t_i) \in S_{i-1}[T_{\sigma_i}]$ and $h(T_{\sigma_i}) = \sum_{j=0}^{m+n_i} b_j\, T_{\sigma_i}^j$, with $b_j \in S_{i-1}$.

Distinguish two cases.

1. $m < n_i$. Then $(**)$, with $g(t_i,T_{\sigma_i}) = \sum_{j=0}^{m} a_j(t_i)T_{\sigma_i}^j$ yields :
   $a_0(t_i) = b_{n_i}, \ldots, b_{n_i+m} = a_m(t_i)$. Thus $a_j(t_i) \in S_{i-1}$ for $j = 0,\ldots,m$.
   However also : $t_i\, a_j(t_i) = b_j$, $j \leq m$, but this contradicts $t_i \notin S_{i-1}$.

2. $m \geq n_i$. Comparing coefficients occuring in $(*^**)$ we get three systems of equations :

$$\text{I} \quad \left\{ \begin{array}{l} a_0^*(t_i)t_i = C(t_i)\, b_0 \\ \hline a_{n_i-1}^*(t_i)t_i = C(t_i)b_{n_i-1} \end{array} \right. \qquad \text{II} \quad \left\{ \begin{array}{l} a_{n_i}^*(t_i)t_i + a_0^*(t_i) = C(t_i)\, b_{n_i} \\ \hline a_m^*(t_i)t_i + a_{m-n_i}^*(t_i) = C(t_i)\, b_m \end{array} \right.$$

$$\text{III}\qquad \begin{cases} a^*_{m-n_i+1}(t_i) = C(t_i)\, b_{m+1} \\ \text{-----------------------} \\ a^*_m(t_i) = C(t_i)\, b_{m+n_i} \end{cases}$$

Let $r \geq 0$ be the t_i-degree of $C(t_i)$. As $b_{m+n_i} \neq 0$, the last equation of III implies $\deg a^*_m(t_i) = r$. Combining this with the last equation of II we get : $\deg a^*_{m-n_i}(t_i) = r + 1$. Now if $0 \leq m-n_i \leq n_i-1$ then I yields a contradiction : $a^*_{m-n_i}(t_i)t_i = C(t_i)\, b_{m-n_i}$. Otherwise we draw from II the equation :

$$a^*_{m-2n_i}(t_i) + a^*_{m-n_i}(t_i)t_i = C(t_i)\, b_{m-n_i}\ ;$$

then $\deg a^*_{m-2n_i}(t_i)$ has to be equal to $r + 2$ and, either we keep increasing the degree, or we derive from I a contradiction. Because the t_i-degree of $a^*_j(t_i)$, $j = 0,\dots,m$, is limited, the latter thing takes place.

<u>COROLLARY</u>. The generic algebra $\mathcal{D}_{1/k}(G)$ is a skew-field.

<u>THEOREM</u> 91. The exponent of $S_{1/k}(G)$ in its Brauer group $\mathcal{B}r(k(t))$ is equal to the exponent of the abelian group G.

<u>PROOF</u>. Let G be of order $n = p_1^{s_1}\dots p_q^{s_q}$, p_i prime. Denote the center $k(t)$ of $S_{1/k}(G)$ by K. By a well-known theorem, cf. [14] , $S_{1/k}(G) \cong S_1 \underset{K}{\otimes}\dots\otimes S_q$, where S_j is of degree $p_j^{s_j}$. Starting from the cyclic decomposition $G \cong \prod_{i=1}^r \mathbb{Z}/n_i\mathbb{Z}$, we compute the p_j-component of G. If $n_i = p_1^{s_{i1}}\dots p_q^{s_{iq}}$, then, for $j = 1,\dots,q$, $s_j = \sum_{i=1}^r s_{ij}$ and the p_j-component G_j of G is $G_j = \prod_{i=1}^r \langle\sigma_i^{n_{ij}}\rangle$ with $n_{ij} = n_i/p_j^{s_{ij}}$ and σ_i being a generate for $\mathbb{Z}/n_i\mathbb{Z}$. Let l_j be the p_j-extension of l, then S_j may be written as $l_j(t)[V_{\sigma_i}^{n_{ij}}, i = 1,\dots,r]$ and the $p_j^{s_{ij}}$-th power of $V_{\sigma_i}^{n_{ij}}$ equals t_i. Hence S_j is an algebra precisely of the type earlier constructed, i.e., $S_j \cong S_{1/k}(G_j) \underset{k_j}{\otimes} k_j(u)$ where k_j is the center of $S_{1/k}(G_j)$ while u

stands for the subset $\{u_1,\ldots,u_m\}$ of $t = \{t_1,\ldots,t_r\}$ consisting of the variables which do not occur as $(V_{\sigma_i}^{n_{ij}})$ exp $p_j^{s_{ij}}$ (p_j does not necessarily divide all n_i).

Write $V_{\sigma_i}^*$ for $V_{\sigma_i}^{n_{ij}}$. Adding u to the center of $S_{l/k}(G_j)$ does not affect the exponent in the corresponding Brauer groups. Therefore, since $e(S_{l/k}(G)) = \prod_{j=1}^{q} e_j$, $e_j = e(S_j)$, it follows that we may assume that G is a p-group from now on. Assume $n = p^m$ and G of type $(p^{m_1},\ldots,p^{m_r})$, i.e., $n_i = p^{m_i}$ and $m = m_1 +\ldots+ m_r$. $S_{l/k}(G)$ decomposes into the product of cyclic algebras, $S_{l/k}(G) = \otimes_{i=1}^{r} S_i$, where each S_i is a cyclic crossed product algebra $(\mathbb{Z}/n_i\mathbb{Z}, l_i(t)/k(t), \{t\}_i)$ which is a division algebra of index p^{m_i}, (l_i/k is the subextension of l/k left fixed by G mod $\mathbb{Z}/n_i\mathbb{Z}$). Up to a permutation we assume $m_1 \geqslant\ldots\geqslant m_r$. If $m_1 =\ldots= m_k$ but $m_{k+1} \neq m_k$ then the hypothesis $e(\otimes_{j=1}^{k} S_j) = p^{m_1}$ yields $e(S_{l/k}(G)) = p^{m_1}$ because for any j, $m_{k+1} \leqslant j \leqslant m_1$ we have

$$S_{l/k}(G)^{p_j} \cong (S_1 \otimes\ldots\otimes S_k)^{p_j} \underset{k(t)}{\otimes} S_{k+1}^{p_j} \otimes\ldots\otimes S_r^{p_j}$$

$$\cong (S_1 \otimes\ldots\otimes S_k)^{p_j} \underset{k(t)}{\otimes} M_N(k(t)), \quad N > 1.$$

The problem is thus reduced to the case where $m_1 =\ldots= m_r$. Using obvious notations for the factor sets we get that $S_{l/k}^{p^d}$, with $d < m_1$ is similar to $(G, l(t)/k(t), \{t_1^{p^d},\ldots,t_r^{p^d}\})$. This means that $S_i^{p^d}$ is similar to $S_i' = (\mathbb{Z}/p^{m_i}\mathbb{Z}, l_i(t)/k(t), t_i^{p^d})$. Write $S_i' = l_i(t)[V_i^*]$, where $(V_i^*)^{n_i} = t_i^{p^d}$ and $V_i^*\lambda = \lambda^{\sigma_i}V_i^*$ for all $\lambda \in l_i(t)$. Put $t_i^{p^d} = s_i$, $s = \{s_1,\ldots,s_r\}$ and look at the subalgebra S_i'' of S_i', $S_i'' = l_i(s)[V_i^*, i = 1\ldots r]$, which is isomorphic to

$$(\mathbb{Z}/p^{m_i}\mathbb{Z}, l_i(s)/k(s), \{s_1,\ldots,s_r\})$$

and thus isomorphic to S_i under the map ϕ_i defined by $\phi_i(t_j) = s_j$, $\phi(V_i) = V_i^*$, (ϕ_i defines an unramified pseudo-place which is an isomorphism on the groundfield, thus ϕ_i is an isomorphism). However, S_i''

being a k(s)-central simple algebra contained in S_i', it follows that $S_i' = S_i'' \otimes_{k(s)} k(t)$ and from this we may derive an isomorphism :

$$S'_{l/k}(G) = \bigotimes_{k(t)} S_i' \cong \bigotimes_{k(t)} (S_i'' \otimes_{k(s)} k(t)) \cong S''_{l/k} \otimes_{k(s)} k(t)$$

with $S''_{l/k}(G) = \bigotimes_{k(s)} = S_i''$. Since $S''_{l/k}(G)$ is isomorphic to $S_{l/k}(G)$ under the map $\phi_1 \otimes \ldots \otimes \phi_r : S_{l/k}(G) \to S''_{l/k}(G)$, which maps l(t) onto $\bigotimes_{k(s)} l_i(s) = l(s)$.

If the k(s)-central division algebra $S''_{l/k}(G)$ were completely split by k(t) then the degree [k(t) : k(s)] would be at least equal to $p^m = [l(s) : k(s)]$. However, $[k(t) : k(s)] = p^{rd} < p^m$ because $d < m_1$.

COROLLARIES. If G is cyclic then $e(S_{l/k}(G))$ is equal to its index. If the order of G is a prime power then $S_{l/k}(G)$ is a primary algebra. One would like to characterize the finite abelian groups which occur as Galois groups of maximal subfields within $S_{l/k}(G)$ or $S_k(G)$. This is unsolved but a first step in this direction is :

PROPOSITION 92. If char k does not divide the order of G, then $S_k(G)$ cannot contain an abelian subfield $F/k_G(t)$ admitting an automorphism of order strictly greater than the exponent G.

The proof is in [17].

V. 3. The Modular Case.

Let k be a field with char $k = p \neq 0$. A purely inseparable extension P of k is called a modular extension of k, if k and P^{p^i} are linearly disjoint over $k \cap P^{p^i}$ for all positive integers i. A finite purely inseparable extension P/k is said to have a basic group G if there exist a k-basis B for P, and a bijective map $G \to B$, $\sigma \to u_\sigma$, such

that $u_\sigma u_\tau = l(\sigma,\tau)u_{\sigma\tau}$ with $l(\sigma,\tau) \in k^*$. This definition may be generalized to inseparable algebras over fields, cf. [36], but it is not necessarily to go into that here.
From the definition it follows that a basic group is an abelian p-group. Indeed, $u_{\tau\sigma} = l^{-1}(\tau,\sigma)u_\tau u_\sigma = l^{-1}(\tau,\sigma)l(\sigma,\tau)u_{\sigma\tau}$, but $u_{\tau\sigma}$ and $u_{\sigma\tau}$ can only be k-dependent if $\sigma\tau = \tau\sigma$. From $u_\sigma u_1 = l(\sigma,1)u_\sigma$ with $l(\sigma,1) \in k^*$ we deduce $u_1 = l(\sigma,1) \in k^*$. Furthermore, the exponent $e(G)$ of G is p^e, and e is the exponent of the extension P/k; indeed, since $\sigma^{e(G)} = 1$ for every $\sigma \in G$ we obtain that $u_\sigma^{e(G)} = \lambda u_1$ with $\lambda \in k^*$, hence $u_\sigma^{e(G)} \in k^*$ and $p^e | e(G)$ follows. Vice versa, $u_\sigma^{p^e} \in k^*$ for all $\sigma \in G$ implies $u_\sigma^{p^e} = \mu\, u_{\sigma p^e}$ with $\mu \in k^*$, thus $u_{\sigma p^e} \in k^*$ or $u_{\sigma p^e} = \lambda\, u_1$ with $\lambda \in k^*$, entailing $\sigma p^e = 1$ and $e(G) | p^e$.

<u>PROPOSITION</u> 93. A finite purely inseparable extension P/k is modular if and only if it has a basic group.

<u>PROOF</u>. It is possible to show that P/k has a basic group if and only if it is a regular extension in the sense of [15]. Also, P/k has a basic group if and only if P/k is isomorphic to the tensor product over k of simple purely inseparable subextensions of P/k. These simple factors correspond to the cyclic subgroups of G occuring in a fixed cyclic decomposition of G, cf. [36].

<u>THEOREM</u> 94. Let P/k have basic group G. Then G is up to isomorphism uniquely determined by the extension P/k.

<u>PROOF</u>. Induction on the exponent e of P/k. If $e = 1$ and $[P : k] = p^n$, then clearly, every basic group of P/k has to be of type $(p,p,\ldots,p)$. Let $e > 1$ and let G and G' be basic groups for P/k with associated k-bases $\{u_\sigma, \sigma \in g\}$ and $\{v_\tau, \tau \in G'\}$ resp. It is immediate that

$F = k[u_\sigma^p, \sigma \in G]$ and $F' = k[v_\tau^p, \tau \in G']$ coincide because v_τ^p may be expressed in the u_σ^p with coefficients in k, and vice versa. The maps $f : G^p \to F$, $f(\sigma^p) = u_\sigma^p$ and $f' : G'^p \to F$, $f'(\tau^p) = v_{\tau^p}$, give rise to basic groups G^p and G'^p for F/k. The exponent of F/k is $e - 1 \geqslant 1$ and by the induction hypothesis $G^p \cong G'^p$. Hence the type of G and G' can only differ in the number of times a factor p occurs in the type; the fact that both groups have the same order p^n implies $G' \cong G$.

A k-algebra N will be called a Galois algebra with group G if N is the direct sum of isomorphic fields, $N = N_1 \oplus \ldots \oplus N_m$, such that G is a group of k-automorphisms with the following properties.

1. The unique element of G leaving N_i fixed is the unit element.
2. G acts transitive on $\{N_1, \ldots, N_m\}$.
3. Any $x \in N$ such that $x^\sigma = x$ for every $\sigma \in G$ is in k.

It is easily seen that N/k is a Galois algebra with group G if and only if there exists a normal k-basis $\{b^\sigma, \sigma \in G\}$ for N. In [15] it is shown that, if A/k is a k-central simple algebra of dimension $[P : k]^2$ where P/k is a modular extension, then P splits A if and only if $A = (G, N/k, \{C_{\sigma,\tau}\})$ for some Galois algebra N with group G and symmetric (k-rational) factor set $\{C_{\sigma,\tau}\}$. The crossed product structure is defined as in the case where N is a field. Let G act in $k(X) = k(\{X_\sigma, \sigma \in G\})$ as follows, $\tau X_\sigma = X_{\tau\sigma}$, $\tau, \sigma \in G$. The symmetric group S_n acts in $k(X)$ by permuting the variables X_σ. The subfield of $k(X)$ left fixed for the action of S_n is $k_S = k(s_1, \ldots, s_n)$, the field generated over k by the symmetric functions in the variables X_σ, $\sigma \in G$. The set $\{s_1, \ldots, s_n\}$ is k-algebraic independent.

<u>PROPOSITION</u> 95. Let N/k be a Galois algebra with finite abelian group G. There exists a k-pseudo-place ψ of $k(X)/k_S$ with residue algebra N/k and $\dim \psi|k = 0$, which is compatible with Galois action.

PROOF. Let k_G be the subfield of $k(X)$ left fixed for the action of G. Consider $f = \prod_{\sigma \in G} (X - X_\sigma) \in k_G[X]$. This is an irreducible polynomial over k_G but also over k_S. The $b^\sigma \in N$, all $\sigma \in G$, $b \in N$, are roots of a polynomial $\bar{f} = \prod_{\sigma \in G} (X - b^\sigma) = \sum_{i=0}^{n} a_i X^{n-i}$ with $a_i \in k$.
The specialization $s_i \longmapsto a_i$ extends to a k-place ϕ of k_S with residue field k. Let O_S be the valuation ring of ϕ and consider $O_S[\{X_\sigma, \sigma \in G\}] = k(X)'$. We have $k(X)' \cap k_S = O_S$. Extend ϕ to a map of $k(X)'$ onto N putting $\phi(X_\sigma) = b^\sigma$. This is easily seen to be a homomorphism, by induction, as follows. First extend ϕ to $O_S[X_1]$; then $\phi_1 : O_S[X_1] \to k[b_1]$ is a homomorphism since X_1 satisfies $f(X_1) = 0$ and b_1 satisfies $\bar{f}(b_1) = 0$, which is the equation obtained by reduction of $f(X_1) = 0$ under ϕ. Let $\phi_1, \ldots, \phi_{i-1}$ be obtained this way. Then X_i (i stands for σ_i) satisfies $f_i(X_i) = (f / \prod_{j=1}^{i-1} (X - X_j))(X_i) = 0$, where f_i is $k_S[X_1, \ldots, X_{i-1}]$-irreducible. This polynomial reduces to $\bar{f}_i = \bar{f} / \prod_{j=1}^{i-1} (X - b_j)$ with coefficients in $k[b_1, \ldots, b_{i-1}]$, ($b_j = b^{\sigma_j}$). So ϕ_{i-1} extends to a homomorphism of $O_S[X_1, \ldots, X_i]$ onto $k[b_1, \ldots, b_i]$.

The following theorem yields a generic description of the p-component of $\mathcal{Br}(k)$.

THEOREM 96. For any class α in the p-component of $\mathcal{Br}(k)$ there exists a representative $A \in \alpha$ and a finite abelian p-group G such that :

1. $A \cong (G, N/k, \{C_{\sigma,\tau}\})$, where N is a Galois algebra with group G and $\{C_{\sigma,\tau}\}$ a symmetric factor set.
2. A/k is residue algebra of $\mathcal{D}_k(G)/k_S(t)$ under a k-pseudo-place of dimension 0 which is compatible with Galois action.

PROOF. Any central simple algebra representing α is a p-algebra and hence it has a purely inseparable splitting field P'/k, say. Let P/k be the modular closure of P'/k (cf. [33]) and let G be the, up to

isomorphism, unique p-group determined by P/k. There exists an $A \in \alpha$ such that $[A : k] = [P : k]^2$, hence, since P splits A we may conclude that $A \cong (G, N/k, \{C_{\sigma,\tau}\})$ for some Galois algebra N/k with group G and symmetric factor set $\{C_{\sigma,\tau}\}$.
Put $A = N[u_\sigma, \sigma \in G]$ with $u_\sigma\lambda = \lambda^\sigma u_\sigma$ for all $\lambda \in N$. The pseudo-place $(k(X)', \phi, N/k)$ constructed in the foregoing proposition extends to a pseudo-place ψ of $\mathcal{D}_k(G)/k_S(t)$ in the obvious way. Since the residue field of $k_S(t)$ under ϕ is exactly k, we have that ψ has k-dimension equal to zero.
In general, when N is not a field, it is impossible to extend ϕ to an unramified pseudo-place of $\mathcal{D}_k(G)$ over $k_G(t)$.

<u>References for Section</u> V.

A.A. ALBERT [1]; I.N. HERSTEIN [14]; K. HOECHSMAN [15]; W. KUYK, P. MULLENDER [16]; W. KUYK [17]; M.E. SWEEDLER [33]; F. VAN OYSTAEYEN [35], [36].

VI. APPENDIX : LOCALIZATION OF AZUMAYA ALGEBRAS

VI. 1. The Center of $Q_\sigma(R)$.

If R is a ring with unit then its centroid is a commutative ring coinciding with the center of R, it will be denoted by C throughout. For an arbitrary idempotent kernel functor σ on M(R), the center of $Q_\sigma(R)$ will be denoted by C_σ. Let σ_0 be the symmetric kernel functor on M(R) defined by the filter T_0 = {left ideals of R containing an ideal A of R for which, Ax = 0 with $x \in R$ implies x = 0}. Although some results in this section may be stated in a slightly generalized form, we will always consider symmetric kernel functors σ on M(R) such that $\sigma \leq \sigma_0$, i.e., such that R is σ-torsion free. By definition of $Q_\sigma(R)$, an $\alpha \in Q_\sigma(R)$ may be represented by an R-linear map $\alpha : A \to R$ where A is an ideal in $T(\sigma)$. The element of $Q_\sigma(R)$ and an R-linear map representing it will be denoted by the same symbol. It is easily seen that an R-linear map representing $0 \in Q_\sigma(R)$ has to be the zero homomorphism. Consequently, if $\alpha : A \to R$ and $\beta : B \to R$ represent the same element of $Q_\sigma(R)$ then α and β coincide on $A \cap B$. Elements of C_σ are characterized in the following proposition.

PROPOSITION 97. 1. If $\alpha \in Q_\sigma(R)$ then $\alpha \in C_\sigma$ if and only if $r\alpha = \alpha r$ for all $r \in j_\sigma(R)$, where j_σ is the canonical ring homomorphism $R \to Q_\sigma(R)$.

2. Let $A \in T(\sigma)$ be an ideal and let $\alpha : A \to R$ represent $\alpha \in Q_\sigma(R)$. If α is left and right R-linear then $\alpha \in C_\sigma$.

3. If $\alpha \in C_\sigma$ then every representative $\alpha : A \to R$ such that A is an ideal in $T(\sigma)$ is left and right R-linear.

PROOF. 1. Right multiplication by α, $m_\alpha : R \to Q_\sigma(R)$, is R-linear and thus m_α extends in a unique way to a left R-linear map

$$\tilde{m}_\alpha : Q_\sigma(R) \to Q_\sigma(R).$$

Since $r\alpha = \alpha r$ for all $r \in j_\sigma(R)$ it follows that left multiplication by α is left R-linear and its restriction to R coincides with m_α. The uniqueness of $\tilde{m}_\alpha$ implies that $\alpha q = q\alpha$ for all $q \in Q_\sigma(R)$ and thus $\alpha \in C_\sigma$.

3. Let $a \in A$, $x \in R$. Then $\alpha(ax) = \alpha\, m_x(a) = m_x\, \alpha(a)$ because j_σ is defined by $j_\sigma(x) = m_x$. Hence $\alpha(ax) = m_x\, \alpha(a) = \alpha(a)x$, proving that α is right R-linear.

2. Suppose that $\alpha : A \to R$ is left and right R-linear and let $\beta : B \to R$ be any left R-linear map with $B \in T(\sigma)$. Then $\alpha\beta$ and $\beta\alpha$ are defined on $AB \in T(\sigma)$ and for all $a \in A$, $b \in B$ we have that :

$$\alpha\beta(ab) = \alpha(a\beta(b)) = \alpha(a)\beta(b)$$

$$\beta\alpha(ab) = \beta(\alpha(a)b) = \alpha(a)\beta(b).$$

Thus $\alpha\beta$ and $\beta\alpha$ coincide on $AB \in T(\sigma)$ and therefore they represent the same element of $Q_\sigma(R)$. Note that $AB \neq (0)$ because $T(\sigma) \subset T_0$.

LEMMA 98. If $\sigma_0 \geq \tau \geq \sigma$ then there is a ring homomorphism $C_\sigma \to C_\tau$.

PROOF. Let $\alpha : A \to R$ represent $\alpha \in C_\sigma$ and let $\beta : B \to R$ with $B \in T(\tau)$ represent $\beta \in Q_\tau(R)$. Then $\alpha\beta$ and $\beta\alpha$ coincide on $AB \in T(\tau)$ and hence they define the same element of $Q_\tau(R)$. This proves that the τ-class of α is in C_τ. Hence, the restriction of the canonical ring homomorphism $Q_\sigma(R) \to Q_\tau(R)$ maps C_σ in C_τ. Since R is τ-torsion free, this map is injective.

Note that the fact that $Q_\sigma(R) \to Q_\tau(R)$ is a ring homomorphism is not a direct consequence of Theorem 10 but it follows because $R/\tau(R)$

and $R/\sigma(R)$ coincide with R.

If R is a prime ring then $\sigma_0 = \sigma^*$ and then we have injections $C_\sigma \to C_*$, and $C_\sigma = C_* \cap Q_\sigma(R)$ for every symmetric σ. It is well-known that C_* is a field, thus if R is prime then the rings C_σ are integral domains.

PROPOSITION 99. Let R be a semiprime ring and let $\alpha \in C_\sigma$ be a zero-divisor of C_σ. Then at least one (and thus all) of its representatives $\alpha : A \to R$, where A is an ideal in $T(\sigma)$, is not a monomorphism.

PROOF. Let $\rho(\sigma,0)$ be the unique R-linear map $Q_\sigma(R) \to Q_0(R)$ extending the identity of R, i.e., $\rho(\sigma,0)[A,\alpha]_\sigma = [A,\alpha]_0$. Since $Q_\sigma(R)$ is σ_0-torsion free it follows that we have inclusions :

$$R \xrightarrow{j_\sigma} Q_\sigma(R) \xrightarrow{\rho(\sigma,0)} Q_0(R) .$$

From Lemma 98 we derive that $\rho(\sigma,0)$ maps C_σ into C_0 and zero-divisors of C_σ map onto zero-divisors in C_0. Let $\alpha : A \to R$ represent a zero-divisor $\alpha \in C_\sigma$, then there does not exist an ideal $B \in T_0$, $B \subset A$, such that $\alpha|B$ is injective, cf. [A.1], because $\rho(\sigma,0)\alpha$ is a zero-divisor in C_σ. If $\alpha' : A' \to R$ were an injective representative for α then $\alpha|A \cap A' = \alpha'|A \cap A'$ is injective and then the fact that $A \cap A' \in T(\sigma) \subset T_0$ yields a contradiction.

Remark. The converse of the above proposition holds in case $\sigma = \sigma_0$ or also for arbitrary $\sigma \leq \sigma_0$ if every ideal in R contains a central element. Indeed, if $\rho(\sigma,0)(\alpha)\beta = 0$ for some $\beta \in C_0$ then $\rho(\sigma,0)(\alpha)I\beta = 0$ for some ideal $I \in T_0$ such that $0 \neq I\beta \subset R$. Hence $0 = I\beta\, \rho(\sigma,0)(\alpha) = \rho(\sigma,0)(I\beta\alpha)$, entailing $I\beta\alpha = 0$. Now, $I\beta$ is an ideal of R because β commutes with R in $Q_0(R)$, and thus,

if $\gamma \in C \cap I\beta$, $\gamma \neq 0$, then $\gamma\alpha = 0$ yields that α is a zero-divisor in C_σ.
Therefore, if some $\alpha : A \to R$ representing $\alpha \in C_\sigma$ is not injective then $\rho(\sigma,0)(\alpha)$ is also represented by $\alpha : A \to R$ and it is then a zero-divisor in C_0. The foregoing then proves the converse to Proposition 99.

Generalizing a result of S.A. Amitsur concerning σ_0, cf. [A.1], we obtain :

PROPOSITION 100. If $R\,C_\sigma$ is semisimple Artinian, then $Q_\sigma(R) = R\,C_\sigma$, σ is a T-functor and an element $a \in R$ is regular if and only if $Ra \in T(\sigma)$.

PROOF. Let A be an ideal in $T(\sigma)$ and $e \in R\,C_\sigma$ a non-zero idempotent, then $A \cap eAe \neq (0)$. Indeed, pick $B \in T(\sigma)$ such that $Be \subset R$. Write $e = \sum_i{}' r_i c_i$ with $r_i \in R$, $c_i \in C_\sigma$ and pick $D \in T(\sigma)$ such that $Dc_i \subset R$ for all i; then $eD \subset R$. Take $I \in T(\sigma)$, $I \subset B \cap D$ and take I to be an ideal. Then $eIAIe \subset eAe \cap A$ and if $eIAIe$ were (0) then $xAx = 0$ for any $x \in IeI$. Since $R\,C_\sigma$ is semiprime, R is semiprime too and $Ax = 0$ follows. However $x \in Q_\sigma(R)$ yields $x = 0$ entailing that $IeI = 0$. Again, this implies $Ie = 0$, $e = 0$, so we reach a contradiction. Let $1 = \Sigma_j\, e_j$ be a decomposition of 1 in $R\,C_\sigma$ and let $a_j \neq 0$ be an element of $A \cap e_jAe_j$. Consider $a = \Sigma_j a_j \in A$. This element is regular in R, for if $ba = 0$ then $ba_j = 0$ and since $e_ja_j = a_j$ we get that $e_jR\,C_\sigma b\, e_j\, a_j = 0$. We know that $e_jR\,C_\sigma e_j$ is a division algebra, thus $e_jR\,C_\sigma b\, e_j = 0$ follows. This entails that $b\, e_j = 0$ because $R\,C_\sigma$ is semiprime, hence $b.1 = 0$ follows. If a is regular in R then it is regular in $R\,C_\sigma$ because $R\,C_\sigma/R$ is σ-torsion where as $R\,C_\sigma$ is σ-torsion free. Therefore a regular element in R is invertible in $R\,C_\sigma$. Every $A \in T(\sigma)$ contains a regular element, thus $1 \in R\,C_\sigma A$ and $1 \in Q_\sigma(R)A$, proving that σ has property (T). Further, if $x \in Q_\sigma(R)$ then $Dx \subset R$ for some $D \in T(\sigma)$, hence $x \in R\,C_\sigma Dx = R\,C_\sigma x$ but $R\,C_\sigma Dx \subset R\,C_\sigma R = R\,C_\sigma$ implies then that $Q_\sigma(R) = R\,C_\sigma$. Finally, if a is regular in R, then $a^{-1} \in Q_\sigma(R)$ and $Da^{-1} \subset R$ for some $D \in T(\sigma)$. Hence $D \subset Ra$ and $Ra \in T(\sigma)$.

Remark. If in Proposition 100, R is also a prime ring, then $Q_\sigma(R)$ is a prime ring too and hence it is a simple Artinian ring. Therefore, for every $0 \neq A \in T(\sigma^*)$ we have that $Q_\sigma(R)A = Q_\sigma(R)$. Property (T) for σ then yields $A \in T(\sigma)$ and $\sigma = \sigma^*$ follows. For prime rings the assumptions made in Proposition 100 force us into the case considered by S.A. Amitsur in [A.1]. This is the reason why we look for milder conditions on R under which $Q_\sigma(R)$ is a central extension of R. This search will lead to the consideration of rings satisfying non-trivial polynomial identities. In the torsion free case R and $Q_\sigma(R)$ will satisfy the same identities and the nature of the identities does not interfer with the localization theory. That is why we present a general approach using Azumaya algebras.

Let R be a ring with unit, C (in) the center of R. Let R^0 be the opposite ring of R, then $R^e = R \otimes_C R^0$ is called the envelopping algebra of R over C. The C-algebra R is said to be separable if and only if R is a projective R^e-module. Let $m : R^e \to \mathrm{Hom}_C(R,R)$ be the ring homomorphism defined by $m(x \otimes y^0)z = xzy$ for all $x,y,z \in R$. Recall from [A.2] the following :

LEMMA 101. Equivalently :

1. R is a separable C-algebra.

2. Put $M = \mathrm{Hom}_{R^e}(R,R^e)$, then $R^eM = R^e$
3. The map $m : R^e \to \mathrm{Hom}_C(R,R)$ is an isomorphism and R is a finitely generated projective C-module.
4. The map m is an isomorphism and C is a direct summand of R as a C-module.

<u>DEFINITION</u>. If R is separable over its center C then R is said to be an <u>Azumaya algebra</u>.

<u>LEMMA</u> 102. The following statements are equivalent :

1. R is an Azumaya algebra.
2. The functors $N \to R \underset{C}{\otimes} N$ and $M \to M^R$, where

   $$M^R = \{m \in M,\ mr = rm \text{ for all } r \in R\},$$

   define an equivalence between $M(C)$ and $M(R^e)$, i.e., between C-modules and R-bimodules.
3. R is a faithful C-module of finite type and $R|mR$ is central simple and finite dimensional for every maximal ideal m of C.

For the proof and more details one may consult [A.2], [A.3]. An extensive and self contained treatment of the theory of Azumaya algebras may also be found in [A.5].

Let R be central separable over C, then there is a one-to-one correspondence between ideals A of R and ideals A^c of C, the correspondence is given by $A^c = A \cap C \leftrightarrow R\,A^c = A$. This follows immediately from Lemma 102, 2.. Let $\sigma \leqslant \sigma_0$ be symmetric on $M(R)$. Then $\{A \cap C,\ A \in T(\sigma)\}$ is the filter of a symmetric functor σ' on $M(C)$ because $C \cap AB = (A \cap C)(B \cap C)$ for all ideals A,B of R. Consider R-modules as C-modules via $C \to R$. Then the functors σ' and σ coincide on R-modules.

<u>LEMMA</u> 103. If R is an Azumaya algebra then there is a natural ring homomorphism $g_\sigma : C_\sigma \to Q_{\sigma'}(C)$.

<u>PROOF</u>. Take an $\alpha \in C_\sigma$ and let $\alpha : A \to R$ be a bimodule homomorphism representing it, A is and ideal in $T(\sigma)$. The restriction $\alpha^c : A^c \to R$ maps A^c into C as is easily checked. Therefore α^c is a C-bimodule morphism $A^c \to C$ and as such, α^c represents a unique element of $Q_{\sigma'}(C)$. All this is cleary independent of the choosen representatives. The correspondence $\alpha \to \alpha^c$ defines thus a map $g_\sigma : C_\sigma \to Q_{\sigma'}(C)$. Moreover, if $\alpha : A \to R$ and $\beta : B \to R$ are bimodule morphisms representing elements of C_σ then $\alpha\beta$ is defined on AB and $(\alpha\beta)^c$ coincides with $\alpha^c\beta^c$ on $(AB)^c = A^cB^c$.

Obviously, an Azumaya algebra R is a prime ring if and only if its center C is an integral domain. In that case C is contained in its field of fractions, K say, and $S = R \otimes_C K$ is a finite dimensional K-central simple algebra.
We now turn to the sheaf theoretic aspect.

<u>PROPOSITION</u> 104. Let R be a prime left Noetherian Azumaya algebra and consider the centers C_A of $Q_A(R)$ where A is an ideal of R. Sticking C_A to the open set $X_A \subset X = \text{Spec } R$ defines a sheaf $\tilde{C}(R)$ of commutative rings on X.

<u>PROOF</u>. If $\sigma_B \geqslant \sigma_A$ then the ring homomorphism $C_A \to C_B$ is the restriction of $\rho(A,B) : Q_A(R) \to Q_B(R)$ and from the fact that $\tilde{Q}$ is a sheaf, we derive directly that $\tilde{C}(R)$ is a mono-presheaf. We are left to prove that for any covering $X_A = \cup X_{A_\alpha}$ (write X_α for X_{A_α}) such that there are $g_\alpha \in C_\alpha$ for which $\rho(\sigma_\alpha, \sup(\sigma_\alpha,\sigma_\beta))g_\alpha = \rho(\sigma_\beta, \sup(\sigma_\alpha,\sigma_\beta))g_\beta$, there is a $g \in C_A$ such that $\rho(\sigma_A,\sigma_\alpha)g = g_\alpha$ for all α. Since $\tilde{Q}$ is a

sheaf, such a g exists in $Q_A(R)$ by Theorem 42. However, since g is mapped into C_α under $\rho(\sigma_A,\sigma_\alpha)$ the injectivity of $\rho(\sigma_A,\sigma_\alpha)$ implies that $g \in C_A = \bigcap_\alpha C_\alpha$.

Remark. In general if R is an Azumaya algebra then there is a Noetherian subring C_0 of C such that $R = R_0 \otimes_{C_0} C$, where R_0 is an Azumaya algebra with center C_0. Thus, as far as purely algebraic questions are concerned, the left Noetherian hypothesis is not very restrictive because one may use general techniques of descent, cf. [A.5]. Geometrically speaking, this assumption is significant because, since Spec is not functorial in general, the problem of relating Spec R to a suitable "patching" of Spec R_0 and Spec C is open.

PROPOSITION 105. Let R be a left Noetherian prime Azumaya algebra with center C and let Spec C be equipped with its classical sheaf. There is a sheaf morphism $\tilde{g}$: Spec C $\to \tilde{C}(R)$ and $\tilde{g}$ is an isomorphism if and only if $\tilde{C}(R)$ is affine.

PROOF. The one-to-one correspondence $A \leftrightarrow A^c$ between ideals of R and ideals of C gives rise to a homeomorphism $\tilde{g}$ of the underlying topological spaces of Spec C and $\tilde{C}(R)$ defined by $\tilde{g}(p) = Rp = P$ for $p \in$ Spec C $= X'$. Since $\tilde{g}^{-1}(X_A) = X'_{A^c}$ we only have to check whether the following diagram of ring homomorphisms is commutative :

$$\begin{array}{ccc} C_A & \xrightarrow{\;g_A\;} & Q_{A^c}(C) \\ {\scriptstyle \rho(A,B)}\downarrow & & \downarrow{\scriptstyle \rho(A^c,B^c)} \\ C_B & \xrightarrow{\;g_B\;} & Q_{B^c}(C) \end{array}$$

for any $B \subset A$, ideals of R.

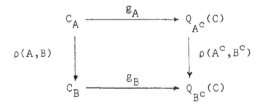

The action of the maps is given as follows : take a representative of the element on which the map has to act and then restrict it to a suitable subset of the domain. Hence commutativity follows easily. Clearly, if $\tilde{C}(R) \cong \text{Spec } C'$ then C' has to be isomorphic to $\bigcap_\sigma C_\sigma = C$ and then $\tilde{g}$ is a sheaf isomorphism.

To end this section, we mention that for an Azumaya algebra R with center C, the zero-divisors of C are characterized by Proposition 99 and its converse, which holds because ideals of R contain central elements.

VI. 2. Localization of Azumaya Algebras.

In this section we drop the assumption $\sigma \leq \sigma_0$. In the sequel R is an Azumaya algebra with center C and $\overline{R} = R/\sigma(R)$, $\overline{C} = C/\sigma(R)$. Moreover, σ will always be a symmetric kernel functor and C_σ is the center of $Q_\sigma(R)$.

THEOREM 106. $Q_\sigma(R)$ is an Azumaya algebra and $Q_\sigma(R) = \overline{R}\, C_\sigma = \overline{R} \otimes_{\overline{C}} C_\sigma$.

PROOF. Since the canonical $R \to \overline{R}$ is surjective it follows that $\overline{R}$ is $\overline{C}$-central separable. Therefore, $\overline{R} \otimes_{\overline{C}} C_\sigma$ is an Azumaya algebra with center C_σ. The ring homomorphism $\overline{R} \otimes_{\overline{C}} C_\sigma \to \overline{R}\, C_\sigma$, defined by multiplication in $Q_\sigma(R)$ is onto and C_σ-linear. Thus $\overline{R}\, C_\sigma$ is an Azumaya algebra with center C_σ. We end up with $\overline{R}\, C_\sigma \subset Q_\sigma(R)$ and both rings having the same center. Hence, $Q_\sigma(R) = \overline{R}\, C_\sigma \otimes_{C_\sigma} D$ (cf. [A.3]) where D is the commuting ring of $\overline{R}\, C_\sigma$ in $Q_\sigma(R)$. Since D commutes with $\overline{R}$ and since Proposition 97.1. obviously holds without $\sigma \leq \sigma_0$, we get $D = C_\sigma$ and $\overline{R}\, C_\sigma = Q_\sigma(R)$. Since $Q_\sigma(R)$ is an $\overline{R}$-bimodule we may use Lemma 102.2. to conclude that $Q_\sigma(R) = \overline{R} \otimes_{\overline{C}} C_\sigma$.

COROLLARIES. 1. An Azumaya algebra R is σ-perfect for every symmetric T-functor σ on $M(R)$.

2. If R is prime and $\sigma = \sigma^*$ then C_σ is a field and $Q_\sigma(R)$, being central separable over a field, is a C_σ-central simple algebra, $Q_\sigma(R) \cong R \underset{C}{\otimes} K$ where K is the field of fractions of C.

3. If R is prime and left Noetherian then we obtain a sequence of sheaf morphisms :

$$\operatorname{Spec} Q_{\sigma'}(C) \xrightarrow[\tilde{g}_\sigma]{} \operatorname{Spec} C_\sigma \xrightarrow[\tilde{g}(\sigma)]{} \tilde{C}(Q_\sigma(R))$$

where $\tilde{g}_\sigma$ derives from $g_\sigma : C_\sigma \to Q_\sigma(C)$ by functoriality of Spec in the commutative case, where as $\tilde{g}(\sigma)$ exists by Proposition 105 if C,R are replaced by $C_\sigma, Q_\sigma(R)$ resp..

<u>THEOREM</u> 107. If σ is a symmetric T-functor then $C_\sigma \cong Q_{\sigma'}(C)$.

<u>PROOF</u>. Consider $\pi : R \to \bar{R}$. Let $\pi\sigma$ be the symmetric kernel functor on $M(\bar{R})$ given by $T(\pi\sigma) = \{\pi(A),\ A \in T(\sigma)\}$. The corollaries to Proposition 33 imply that $\pi\sigma$ is a symmetric T-functor. Moreover, $C/\sigma(R) \cong C/\sigma'(R)$ and $\pi\sigma'$ is a T-functor whenever σ' is a T-functor. Thus, the fact that $Q_\sigma(R) = Q_{\pi\sigma}(\bar{R})$ and $Q_{\sigma'}(C) = Q_{\pi\sigma'}(\bar{C})$ yields that in proving this theorem we may assume that R is σ-torsion free, i.e., $R = \bar{R}$, $\sigma = \pi\sigma \leq \sigma_0$. The inclusion $C \to Q_{\sigma'}(C)$ extends to a unique C-linear $h : C_\sigma \to Q_{\sigma'}(C)$ because C_σ/C is σ'-torsion. When g_σ is restricted to C, it induces the canonical $C \to Q_{\sigma'}(C)$ and this entails that g_σ is injective because g_σ has to coincide with h by the uniqueness of h. Note that this has been derived without using property (T). Let A^c be an ideal of C and put $A = RA^c$. If $A^c \in T(\sigma')$ then $A \in T(\sigma)$ and $Q_\sigma(R)A = Q_\sigma(R)$ by property (T) for σ. Proposition 106 entails, $RC_\sigma A = RC_\sigma A^c = RC_\sigma$, thus $RC_\sigma(C_\sigma A^c) = RC_\sigma$ and since RC_σ is separable over C_σ this means that $C_\sigma A^c = C_\sigma$ and a

fortiori $Q_{\sigma'}(C)A^c = Q_{\sigma'}(C)$. This implies that σ' is a T-functor and because every $A^c \in T(\sigma')$ extends to $C_\sigma A^c = C_\sigma$ this entails that $C_\sigma = Q_{\sigma'}(C)$. Since g_σ is the unique C-linear extension of $C \to Q_{\sigma'}(C)$ to C_σ, it is immediate that g_σ is an isomorphism.

<u>COROLLARY</u>. Let X_A be open in X = Spec R, where R is a prime left Noetherian ring. If X_A is a T-set then we get a commutative diagram of sheaf morphisms :

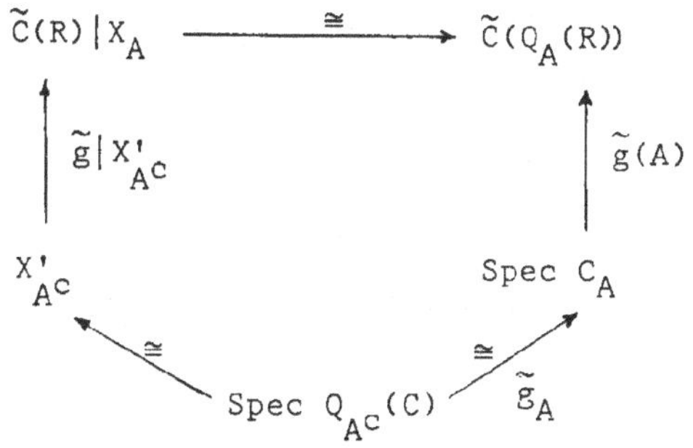

where X' = Spec C. The isomorphisms exist by Theorem 47 and the fact that R is σ_A-perfect. The foregoing proposition yields that $\tilde{g}_A$ is an isomorphism. Commutativity of the diagram is merely verification.

<u>LEMMA</u> 108. If R is an Azumaya algebra then Spec R has a T-basis. Let P be an arbitrary prime ideal of R, then σ_{R-P} is a T-functor.

<u>PROOF</u>. For every $r \in R$ choose $a(r) \in RrR \cap C$ and put $A(r) = Ra(r)$. Then $A(r)$ is an ideal of R and if $B \in T(\sigma_{A(r)})$ then $B \supset A^n(r) = Ra^n(r)$. Moreover, since B contains a B' also in $T(\sigma_{A(r)})$ buth with B' being finitely generated, it follows that $\sigma_{A(r)}$ is a T-functor. In this way a T-basis $\{X_{A(r)}, r \in R\}$ is obtained for Spec R. To prove that σ_{R-P} is a T-functor one argues in a similar way. Note that in the

absence of the left Noetherian condition the lemma deals with Spec as a topological space, therefore the second statement is not a consequence of the first.

COROLLARY. All sheaf morphisms appearing in the diagram of the corollary to Theorem 107 are in fact sheaf isomorphisms.

PROOF of the Corollary. Let X_B be a T-set contained in the T-set X_A then X_{B^c} is a T-set contained in X_{A^c} and we have

$$C_B(Q_A(R)) \cong Q_{B^c}(C_A).$$

The ring homomorphisms $g_B(A) : C_B(Q_A(R)) \to Q_{B^c}(C_A)$ defining $\tilde{g}(A)$ are isomorphisms whenever X_B is a T-set, i.e. for a basis of the Zariski-topology. Hence $\tilde{g}(A)$ is a sheaf isomorphism and the commutativity of the diagram yields the same for $\tilde{g}|X'_{A^c}$.

THEOREM 109. Let P be a prime ideal of R, then :

1. The elements of $G(P)$ map onto invertible elements in $Q_{R-P}(R)$ under the canonical $R \to Q_{R-P}(R)$.
2. If R is left Noetherian then R satisfies the left Ore condition with respect to $G(P)$.
3. If R is left Noetherian then J. Lambek, G. Michler's torsion theory σ_P has property (T) and $\sigma_P = \sigma_{R-P}$.

PROOF. Step 1. Reduction of 1. and 2.

Consider $\pi : R \to R_1 = R/\sigma_{R-P}(R)$ and denote the prime ideal $\pi(P)$ by P_1. Then $\pi\sigma_{R-P} = \sigma_1$ is exactly the kernel functor associated with the m-system R_1-P_1 and, σ_{R-P} and σ_1 coincide on R_1-modules. Lemma 108 yields that σ_{R-P} is a T-functor hence, by Proposition 33, σ_1

is also a T-functor. Denote $\pi(C)$ by C_1. Then R_1 is an Azumaya algebra with center C_1. Theorem 106 entails that $Q_{\sigma_1}(R_1)$ is a central extension of R_1 and thus P_1 extends to an ideal $Q_{\sigma_1}(R_1)P_1$ of $Q_{\sigma_1}(R_1)$. Since $Q_{R-P}(R_1) \cong Q_{\sigma_1}(R_1)$ it follows that for every $M \in M(R_1)$ we have that $Q_{\sigma_1}(M) = Q_{R-P}(M)$. To prove statements 1. and 2. it will be sufficient to prove the analoges for R_1, P_1 and σ_1. Indeed, $\sigma_{R-P}(R) \subset P$, this yields $\pi G(P) \subset G(P_1)$ and thus the fact that $G(P_1)$ consists of elements which are invertible in $Q_{\sigma_1}(R_1)$ yields statement 1. because $R \to Q_{R-P}(R)$ factorizes via π. In case R is left Noetherian, the fact that $\pi G(P) = G(P_1)$ entails that the left Ore condition for R_1 with respect to $G(P_1)$ lifts to the left Ore condition for R with respect to $G(P)$. Consequently in proving 1. and 2. we will assume that R is σ_{R-P}-torsion free.

Step 2. Proof of 1. Property (T) for σ_{R-P} yields an isomorphism $C_{R-P} \cong Q_{C-p}(C)$, where $p = P \cap C$. Since C is commutative it follows that C_{R-P} is a local ring. Thus $Q_{R-P}(R)$ is central separable over a local ring and $P^e = Q_{R-P}(R)P$ is a maximal ideal such that $Q_{R-P}(R/P) \cong Q_{R-P}(R)/P^c$ is a simple algebra. Therefore P^e is the Jacobian radical of $Q_{R-P}(R)$. If $s \in G(P)$ then the image of s in $Q_{R-P}(R/P)$ is regular, thus invertible in $Q_{R-P}(R/P)$, the latter ring being simple Artinian. Suppose $t \in Q_{R-P}(R)$ is such that $1 - st \in P^e$, hence $st \in 1 + P^e$. Now $1 + P^e$ consists of units of $Q_{R-P}(R)$ and this proves that s is (both left and right) invertible in $Q_{R-P}(R)$.

Step 3. Proof of 2. From $(Rs)^e = Q_{R-P}(R)$ it follows that $Rs \in T(\sigma_{R-P})$. Moreover, $\sigma_{R-P} \leqslant \sigma_P$ implies that $Rs \in T(\sigma_P)$ for every $s \in G(P)$ and this yields that $[Rs : r] \cap G(P) \neq \phi$ for all $r \in R$. The latter is equivalent to the left Ore condition for R with respect to $G(P)$.

Step 4. Proof of 3. Look at the following commutative diagram of ring homomorphisms :

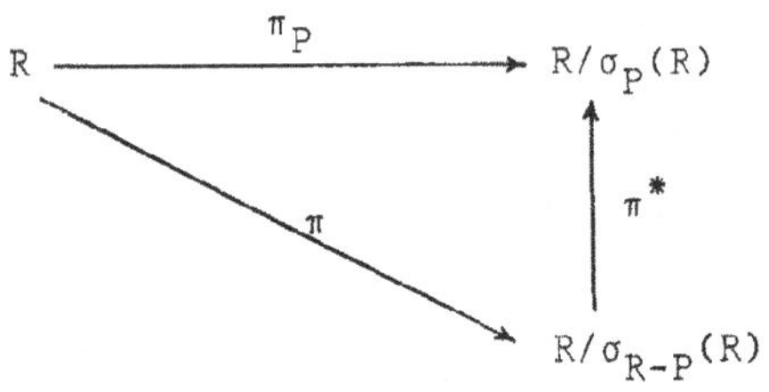

By Proposition 33. 3., π^* extends to a ring homomorphism

$$Q_{R-P}(R) \to Q_{R-P}(R/\sigma_P(R)).$$

Consequently, G(P) maps onto a set of invertible elements in $Q_{R-P}(R/\sigma_P(R))$. Moreover, $R/\sigma_P(R)$ satisfies the left Ore condition with respect to $G(\pi_P(P) = \pi_P(G(P))$.
Thus σ_P has $\{Rs, s \in G(P)\}$ for a filterbasis and thus it follows immediately that σ_P has property (T). Finally since

$$Q_{R-P}(R)s = Q_{R-P}(R) \text{ for all } s \in G(P)$$

we get $Rs \in T(\sigma_{R-P})$ by property (T) for σ_{R-P} and $\sigma_{R-P} = \sigma_P$ follows.

Remark. The canonical $Q_{R-P}(R) \to Q_{\sigma_P}(R)$ is always a ring homomorphism. The proof of this is only a slight extension of Theorem 10. 2.

COROLLARY. Let R be a prime left Noetherian Azumaya algebra, then :

1. Spec R has a T-basis, all stalks are T-stalks and each stalk is a local ring with simple Artinian residue ring.
2. If X_A is a T-set of Spec R then $X_A \cong \text{Spec } Q_A(R)$ and there is a one-to-one correspondence between sub T-sets in X_A and T-sets in

Spec $Q_A(R)$. Since R is σ_A-perfect for every ideal A of R, there is also a one-to-one correspondence between geometric subsets of X_A and geometric open sets in Spec $Q_A(R)$.

VI. 3. Azumaya Algebras over Valuation Rings.

THEOREM 110. Let R be an Azumaya algebra of constant rank (n^2) and suppose that C is a semi-local ring. Then $g_\sigma : Q_\sigma \to Q_{\sigma'}(C)$ is an isomorphism for every symmetric kernel functor σ on M(R).

PROOF. Because R is C-projective of finite type and of constant rank it follows that R is a free C-module. The first part of the proof of Theorem 107, which does not make use of property (T) for σ, yields that g_σ is an injective ring homomorphism. Let

$$\pi : R \to \overline{R} = R/\sigma(R)$$

be the torsion reduction. The fact that R is C-free yields that $\text{Ker } \pi \cong \oplus^{n^2} \sigma'(C)$ and thus $\overline{R} \cong \oplus^{n^2} \overline{C}$ where $\overline{C} = C/\sigma(R) = C/\sigma'(C)$. Hence we may assume that we are in the torsion free case $R = \overline{R}$. Recall that g_σ maps $\alpha \in C_\sigma$ represented by $\alpha : A \to R$ where A is an ideal in $T(\sigma)$, onto α^c represented by $\alpha|A^c : A^c \to C$. To prove that g_σ is onto, take $\beta \in Q_{\sigma'}(C)$ and let $\beta : B \to C$ be a representative for β with $B \in T(\sigma')$. If R is written $Cx_1 \oplus \ldots \oplus Cx_{n^2}$, then $A = RB$ is equal to $Bx_1 \oplus \ldots \oplus Bx_{n^2}$.
Define $\alpha : A \to R$ by

$$\alpha\left(\sum_{i=1}^{n^2} b_i x_i\right) = \sum_{i=1}^{n^2} \beta(b_i) x_i$$

where $b_i \in B$. It is readily verified that α is left and right R-linear. Therefore α represents an element $\alpha \in C_\sigma$ which is mapped onto $\alpha^c = \beta$ under g_σ.

COROLLARY. If C is a local ring then $C_\sigma \cong Q_{\sigma'}(C)$ for every symmetric σ. Indeed over a local ring, every projective module of finite type is free, thus the condition about the rank being constant may be dropped in that case.

PROPOSITION 111. If R is an Azumaya algebra over a valuation ring, then there is a one-to-one correspondence between symmetric kernel functors σ on $M(R)$ and prime ideals P of R, i.e., $\sigma = \sigma_{R-P}$. Thus every symmetric kernel functor has property (T).

PROOF. The prime ideals of C are linearly ordered so $C(\sigma')$ consists of one element, p say. Now C is commutative and p is a prime ideal of C, therefore localization at $C-p$ has property (T) and thus σ' is a T-functor. Theorem 110 implies that σ is a T-functor if and only if σ' is a T-functor because of the one-to-one correspondence between ideals of $Q_\sigma(R)$ and ideals of C_σ, and between ideals of R and ideals of C. Put $P = Rp$. Every ideal A of R not contained in P, contains an element $s \in C-p$. But then s is invertible in C_σ and $Q_\sigma(R)A = Q_\sigma(R)$ follows. This yields $A \in T(\sigma)$ by property (T) for σ, thus this entails $\sigma = \sigma_{R-P}$.

These results provide the link between pseudo-places of simple algebras and localization of Azumaya algebras. Let K be a field, let C be a valuation ring of K and denote C by O_K in what follows. Let M_K be the radical of O_K and let ϕ be the place of K with valuation ring O_K and residue field k.

PROPOSITION 112. Let $\mathcal{D}$ be a K-algebra and let $(R,\psi,\mathcal{D}_1/k)$ be a ϕ-pseudo-place of $\mathcal{D}/K$ such that $\mathcal{D}_1 = R/RM_K$ is k-central simple. Then,

1. For every field l such that there exist places : ϕ_1 of K with residue field l and ϕ_2 of l with residue field k, there is a symmetric T-functor σ such that there exists a pseudo-place $(Q_\sigma(R),\Omega_\sigma,R_\sigma/l)$ for which R_σ is an l-central simple algebra.
2. For every field l as in 1., $(Q_\sigma(R),\Omega_\sigma,R_\sigma/l)$ is an unramified pseudo-place of $\mathcal{D}/K$.

PROOF. 1. Since $\mathcal{D}_1$ is simple and O_K is local we have that R is central separable over O_K and then Theorem 106 and Proposition 111 imply that $Q_\sigma(R)$ is central separable over $O_\sigma = p^{-1}O_K$ for the prime ideal p of O_K corresponding to σ. Choose σ such that O_σ has residue field l, then $Q_\sigma(R)/Q_\sigma(P)$ (with $P = Rp$) is simple because σ has property (T). Hence $\Omega_\sigma : Q_\sigma(R) \to Q_\sigma(R/P) = R_\sigma$ yields the desired pseudo-place of $\mathcal{D}/K$.

2. From the fact that $Q_\sigma(R)$ is an Azumaya algebra with center O_σ we deduce that $[R_\sigma : l] = [Q_\sigma(R) : O_\sigma]$ and then

$$[Q_\sigma(R) : O_\sigma] \geq [\mathcal{D} : K] \geq [R_\sigma : l]$$

proves the equality $[\mathcal{D} : K] = [R_\sigma : l]$ and thus Ω_σ is unramified.

COROLLARY. If $(R,\psi,\mathcal{D}_1/k)$ is a pseudo-place of $\mathcal{D}/K$ such that $\mathcal{D}_1$ is k-central simple then equivalently :

1. ψ is unramified.

2. R is an Azumaya algebra.

Indeed if ψ is unramified then $\mathrm{Ker}\ \psi = RM_K$ and the fact that $\psi(R)$ is simple and k-central yields that R is central separable over O_K. The converse follows from Proposition 112. 2. if one takes $l = k$, $\sigma = \sigma_{R-P}$ where $P = \mathrm{Ker}\ \psi$.

This corollary extends Theorem 56 and Proposition 64. It also has consequences for the theory of primes in skew-fields. Let $(R,\psi,\mathcal{D}_1/k)$ be an unramified pseudo-place of $\mathcal{D}/K$ such that $\mathcal{D}_1$ is a central skew-field.

PROPOSITION 113. Each of the following equivalent statements :

1. P is a completely prime ideal of R
2. $Q_\sigma(P)$ is a completely prime ideal of $Q_\sigma(R)$.
3. R_σ is a skew-field.
4. Ψ is a specialization of Ω_σ as a pseudo-place of $\mathcal{D}/K$.

implies that σ is a prime kernel functor.

PROOF. $1 \Leftrightarrow 2$. If $x,y \in R$ and $xy \in P$ then $xy \in Q_\sigma(P)$ and thus Ix or Iy is contained in P for some $I \in T(\sigma)$. Since $I \not\subset P$ because $P \notin T(\sigma)$, x or y has to be in P, so $2 \Rightarrow 1$. Conversely, if $xy \in Q_\sigma(P)$ with $x,y \in Q_\sigma(R)$ then let I_1, $I_2 \in T(\sigma)$ such that I_1x and I_2y are contained in R. Write I_1^c, I_2^c for $I_1 \cap O_K$, $I_2 \cap O_K$ resp. Then $(I_1I_2)^c xy = I_1^c x I_2^c y \subset R \cap Q_\sigma(P) = P$. If $I_2^c y \not\subset P$ then $cy \notin P$ for some $c \in I_2^c$ and thus the fact that P is completely prime in R yields $I_1^c x \subset P$, entailing that $I_1x = RI_1^c x \subset P$ and thus $x \in Q_\sigma(P)$ by Lemma 13. The equivalence of statements 2. and 3. is obvious. Since Ω_σ is unramified, the implication $4 \Rightarrow 3$ is an immediate consequence of Proposition 62 and Proposition 65. 2. Finally, suppose that $x \in Q_\sigma(P) - \operatorname{Ker} \psi$. Then $Ix \subset P$ for some ideal I in $T(\sigma)$. If $\mathcal{D}_1$ is a skew-field then the fact that Ψ is unramified implies that R is a "valuation" ring of the skew-field $\mathcal{D}$, i.e., $x \notin R$ implies $x^{-1} \in R$. Firstly, if $x \notin R$ then $x^{-1} \in R$ implies that $Q_\sigma(P) = Q_\sigma(R)$ contradicting $P \notin T(\sigma)$.
Secondly, if $x \in R$ then $Ix \subset P$ yields $x \in P$ contrary to the

hypothesis $x \notin \operatorname{Ker} \Psi$ since $P \subset \operatorname{Ker} \Psi = RM_K$. Thus we have

$$Q_\sigma(R) \supset R \supset RM_K \supset P = Q_\sigma(P)$$

and this means that $\Omega_\sigma \to \Psi$ by the theory of specialization of pseudo-places.

We are left to prove that any of the equivalent conditions implies that σ_P is a prime kernel functor. Combining Theorem 109 with Proposition 26. 3. we find that σ_P is a restricted kernel functor, thus $P \subset A$ for every $A \in C(\sigma_P)$. Since R/A is σ_P-torsion free it is clear that $\sigma_P \leq \tau_{R/A}$. On the other hand, if $\bar{x} \in R/P$ is such that $Ix \subset P$ for some $I \in T(\tau_{R/A})$ and a representative $x \in \bar{x}$, then $x \notin P$ would entail $I \subset P$ because P is completely prime, but then $A \supset P$ would also imply that $A \in T(\tau_{R/A})$ which is a contradiction. Therefore R/P is $\tau_{R/A}$-torsion free and $\sigma_P = \tau_{R/P} = \tau_{R/A}$ follows (for every $A \in C'(\sigma)$). Hence σ_P is a prime kernel functor by Proposition 7.

COROLLARY. To every prime $\operatorname{Ker} \Omega$ of the K-algebra $\mathcal{D}$, where Ω is a pre-place of $\mathcal{D}/K$ defined on a subring R_Ω, which specializes to the prime $\operatorname{Ker} \Psi$, there corresponds a prime kernel functor σ on $M(R_\Psi)$ such that $Q_\sigma(R_\Psi) = R_\Omega$.

VI. 4. Modules over Azumaya Algebras.

LEMMA 114. Let R be a left Noetherian Azumaya algebra with center C. Let M,N be R-modules such that M is of finite type where as N has finite representation. If $f : M \to N$ is an R-module homomorphism such that $f_{R-P} : Q_{R-P}(M) \to Q_{R-P}(N)$ is an isomorphism, then there exists a T-functor σ_B such that $f_B : Q_B(M) \to Q_B(N)$ is an isomorphism for some ideal B of R.

PROOF. Since $0 \to f(M) \to N \to N/f(M) \to 0$ is exact it follows that $Q_{R-P}(N/f(M)) = 0$. Hence $N/f(M)$ is σ_{R-P}-torsion and because $N/f(M)$ is of finite type there is an ideal $A \in T(\sigma_{R-P})$ for which $A(N/f(M)) = 0$. Choose $a \in A \cap C$. Then $\sigma_{Ra} = \sigma_a$ is a T-functor such that $f_a : Q_a(M) \to Q_a(N)$ is surjective. Exactness of Q_a implies that $Q_a(M)$ and $Q_a(N)$ are $Q_a(R)$-modules of finite type and $Q_a(N)$ has finite representation. Therefore Ker f_a is of finite type. Since $\sigma_a \leqslant \sigma_{R-P}$, we have $Q_{R-P}(\text{Ker } f_a) = 0$. As before we obtain a T-functor σ_b such that $f_{ba} : Q_b(Q_a(M)) \to Q_b(Q_a(N))$ is an isomorphism. Taking $B = Ra \cap Rb$, then $B \neq (0)$ because $Rab = 0$ would imply that Ra or Rb is in P, and $\sigma_B = \sup\{\sigma_a, \sigma_b\}$. Since $Q_B(R) = Q_b(Q_a(R))$ property (T) yields that f_{ba} is an isomorphism $Q_B(M) \to Q_B(N)$.

Remark. In a situation as in the previous lemma, f is called a <u>local isomorphism</u> because there is a Zariski open set X_B such that $Q_B(M) \cong Q_B(N)$.

THEOREM 115. Let R be a left Noetherian Azumaya algebra and let M be a projective R-module of finite type. Let P be a prime ideal of R and put $p = P \cap C$. Then $M' = Q_{R-P}(M)$ is a free R'-module, where $R' = Q_{R-P}(R)$.

PROOF. From Theorem 109 we derive that $P^e = Q_{R-P}(P)$ is the Jacobson radical of R'. By property (T) for σ_{R-P}, $M' = R' \underset{R}{\otimes} M$ and thus M' is a projective R'-module of finite type, hence M' has finite representation. The equivalence between $M(R^e)$ and $M(C)$ entails that $P^e \cong R' \underset{C'}{\otimes} pC'$ but because C' is commutative and because the action of C' on R' and pC' is the same whether considered on the left or on the right, we have that $P^e \cong pC' \underset{C'}{\otimes} R'$. Consequently,

$$P^e \underset{R'}{\otimes} M' \cong pC' \underset{C'}{\otimes} M'.$$

Now, R' is C'-free of finite rank and M' is projective of finite type, therefore M' is C'-projective of finite type and this entails that $\Psi : pC' \otimes_{C'} M' \to C' \otimes_{C'} M' \cong M'$ deriving from the injection $pC' \to C'$ is also injective. Furthermore,

$$M'/P^eM' \cong Q_{R-P}(R/P) \otimes_{R'} M'$$

is a projective $Q_{R-P}(R/P)$-module of finite type and because $Q_{R-P}(R/P)$ is simple this implies that M'/P^eM' is $Q_{R-P}(R/P)$-free. So we end up with the following situation :

1°. M' has finite presentation as a left R'-module,

2°. The R'-linear $\Psi : P^e \otimes_{R'} M' \to M'$ is injective,

3°. The R'/P^e-module M'/P^eM' is free,

and since P^e is the Jacobson radical of R', this implies that M' is a free R'-module.

PROPOSITION 116. If M is R-projective of finite type then M is locally free. If R is moreover a prime ring then M is locally free of constant rank.

PROOF. Let $F(R)$ be a free R-module of finite rank and let $F(R) \to M \to 0$ be exact. There is an ideal B of R and an associated T-functor σ_B such that $Q_B(F(R)) \cong F(Q_B(R)) \cong Q_B(M)$. To an $M \in M(R)$ there corresponds thus a locally constant function $r : \text{Spec } R \to \mathbb{Z}$ given by $r(P) = [Q_{R-P}(M) : Q_{R-P}(R)]$. If R is a prime ring then any two open subsets of Spec R intersect non-trivially and thus r is constant in that case.

REFERENCES FOR THE APPENDIX

[A.1] S.A. AMITSUR, On Rings of Quotients, Symposia Mathematica, vol. 8, p. 149-164. Academic Press, London 1972

[A.2] M. AUSLANDER, O. GOLDMAN, The Brauer Group of a Commutative Ring, Trans. Americ. Math. Soc., vol. 97 (1960), p. 367

[A.3] G. AZUMAYA, On Maximally Central Algebras, Nagoya Math. J., vol. 2 (1951), p. 119-150

[A.4] I.N. HERSTEIN, Noncommutative Rings, The Carus Mathematical Monographs, number 15, Math. Assoc. of America, 1968.

[A.5] M. KNUS, M. OJANGUREN, Théorie de la Descente et Algèbres d' Azumaya, Lecture Notes in Math. 389, Springer-Verlag, 1974

[A.6] L.W. SMALL, Orders in Artinian Rings, J. of Algebra, vol. 4 (1966), p. 18.

REFERENCES

[1] A.A. ALBERT, New Results on Associative Division Algebras, J. of Algebra 5 (1967), pp. 110-132.

[2] M. AUSLANDER, O. GOLDMAN, Maximal Orders, Trans. Amer. Math. Soc. 97 (1960), pp. 1-24.

[3] G. AZYMAYA, On Maximally Central Algebras, Nagoya Math. J. 2 (1951), pp. 119-150.

[4] A.W. CHATTERS, S.M. GINN, Localization in Hereditary Rings, J. of Algebra 22 (1972), pp. 82-88.

[5] A.W. CHATTERS, A.G. HEINICKE, Localization at a Torsion Theory in Hereditary Noetherian Rings, Proc. London Math. Soc. XXVII, 1973.

[6] I.G.CONNELL, A Natural Transform of the Spec Functor, J. of Algebra 10 (1968), pp. 69-91.

[7] M. DEURING, Algebren, Ergebnisse der Mathematik und ihrer Grenzgebiete, vol. 41, Springer Verlag 1968.

[8] V. DLAB, Rank Theory of Modules, Fund. Math. 64 (1969), pp. 313-324.

[9] P. GABRIEL, Des Catégories Abeliennes, Bull. Soc. Math. France 90 (1962), pp. 323-448.

[10] A.W. GOLDIE, A Note on Non-commutative Localization, J. of Algebra 8 (1968), pp. 41-44.

[11] A.W. GOLDIE, Localization in Non-commutative Noetherian Rings, J. of Algebra 5 (1967), pp. 89-105.

[12] O. GOLDMAN, Rings and Modules of Quotients, J. of Algebra 13 (1969), pp. 10-47.

[13] A.G. HEINICKE, On the Ring of Quotients at a Prime Ideal of a Right Noetherian Ring,
Canad. J. Math. 24 (1972), pp. 703-712.

[14] I.N. HERSTEIN, Noncommutative Rings,
The Carus Math. Monographs 15, Math. Assoc. Amer. 1968.

[15] K. HOECHSMAN, Algebras Split by a Given Purely Inseparable Field,
Proc. Amer. Math. Soc. 14 (1963), pp. 768-776.

[16] W. KUYK, P. MULLENDER, On the Invariants of Finite Abelian Groups,
Indag. Math. 25, Nr. 2, 1963.

[17] W. KUYK, Generic Construction of a Class of Non Cyclic Division Algebras,
J. Pure and Applied Algebra 2 (1972), pp. 121-131.

[18] J. KUZMANOVITCH, Localization of Dedekind Prime Rings,
J. of Algebra 21 (1972), pp. 378-393.

[19] J. LAMBEK, Lectures on Rings and Modules,
Waltham, Toronto, London, 1966.

[20] J. LAMBEK, Torsion Theories, Additive Semantics and Rings of Quotients, Lecture Notes in Mathematics 177 (1971).

[21] J. LAMBEK, G. MICHLER, The Torsion Theory at a Prime Ideal of a Right Noetherian Ring,
J. of Algebra 25, (1973), pp. 364-389.

[22] L. LESIEUR, R. CROISOT, Algèbre Noethérienne Non-commutative,
Mémor. Sci. Math. 154 (1963).

[23] E. MATLIS, Injective Modules over Noetherian Rings,
Pacific J. Math. 8 (1958), pp. 511-528.

[24] D.C. MURDOCH, Contributions to Noncommutative Ideal Theory,
Canad. J. of Math., vol. 6, Nr. 1, 1952.

[25] D.C. MURDOCH, F. VAN OYSTAEYEN, A Note on Reductions of Modules and Kernel Functors, Bull. Math. Soc. Belg., to appear.

[26] D.C. MURDOCH, F. VAN OYSTAEYEN, Noncommutative Localization and Sheaves, J. of Algebra, to appear.

[27] D.C. MURDOCH, F. VAN OYSTAEYEN, Symmetric Kernel Functors and Quasi-primes, Indag. Math., to appear.

[28] M. NAGATA, Local Rings, Wiley-Interscience 1962.

[29] L. SILVER, Noncommutative Localization and Applications, J. of Algebra 7 (1967), pp. 44-76.

[30] S.K. SIM, Prime Idempotent Kernel Functors with Property T, Proc. Amer. Math. Soc., to appear.

[31] S.K. SIM, Noncommutative Localizations and Prime Ideals, Proc. Amer. Math. Soc.

[32] H. STORRER, On Goldman's Primary Decomposition, Lectures on Rings and Modules, Lect. Notes in Mathematics 246, Springer-Verlag, Berlin 1972.

[33] M.E. SWEEDLER, Structure of Inseparable Extensions, Ann. of Math. 87 (1968), p. 401.

[34] F. VAN OYSTAEYEN, On Pseudo-places of Algebras, Bull. Soc. Math. Belg., XXV, 1973, pp. 139.

[35] F. VAN OYSTAEYEN, Generic Division Algebras, Bull. Soc. Math. Belg., XXV, 1973, pp. 259-285.

[36] F. VAN OYSTAEYEN, The p-component of the Brauer Group of a Field of Characteristic $p \neq 0$, Indag. Math. 36, Nr. 1, 1974, pp. 67-76.

[37] F. VAN OYSTAEYEN, Primes in Algebras over Fields, J. Pure Applied Algebra, to appear.

[38] F. VAN OYSTAEYEN, Extension of Ideals under Symmetric Localization, to appear.

SUBJECT INDEX

Vol. 277: Séminaire Banach. Edité par C. Houzel. VII, 229 pages. 1972. DM 20,-

Vol. 278: H. Jacquet, Automorphic Forms on GL(2). Part II. XIII, 142 pages. 1972. DM 16,-

Vol. 279: R. Bott, S. Gitler and I. M. James, Lectures on Algebraic and Differential Topology. V, 174 pages. 1972. DM 18,-

Vol. 280: Conference on the Theory of Ordinary and Partial Differential Equations. Edited by W. N. Everitt and B. D. Sleeman. XV, 367 pages. 1972. DM 26,-

Vol. 281: Coherence in Categories. Edited by S. Mac Lane. VII, 235 pages. 1972. DM 20,-

Vol. 282: W. Klingenberg und P. Flaschel, Riemannsche Hilbertmannigfaltigkeiten. Periodische Geodätische. VII, 211 Seiten. 1972. DM 20,-

Vol. 283: L. Illusie, Complexe Cotangent et Déformations II. VII, 304 pages. 1972. DM 24,-

Vol. 284: P. A. Meyer, Martingales and Stochastic Integrals I. VI, 89 pages. 1972. DM 16,-

Vol. 285: P. de la Harpe, Classical Banach-Lie Algebras and Banach-Lie Groups of Operators in Hilbert Space. III, 160 pages. 1972. DM 16,-

Vol. 286: S. Murakami, On Automorphisms of Siegel Domains. V, 95 pages. 1972. DM 16,-

Vol. 287: Hyperfunctions and Pseudo-Differential Equations. Edited by H. Komatsu. VII, 529 pages. 1973. DM 36,-

Vol. 288: Groupes de Monodromie en Géométrie Algébrique. (SGA 7 I). Dirigé par A. Grothendieck. IX, 523 pages. 1972. DM 50,-

Vol. 289: B. Fuglede, Finely Harmonic Functions. III, 188. 1972. DM 18,-

Vol. 290: D. B. Zagier, Equivariant Pontrjagin Classes and Applications to Orbit Spaces. IX, 130 pages. 1972. DM 16,-

Vol. 291: P. Orlik, Seifert Manifolds. VIII, 155 pages. 1972. DM 16,-

Vol. 292: W. D. Wallis, A. P. Street and J. S. Wallis, Combinatorics: Room Squares, Sum-Free Sets, Hadamard Matrices. V, 508 pages. 1972. DM 50,-

Vol. 293: R. A. DeVore, The Approximation of Continuous Functions by Positive Linear Operators. VIII, 289 pages. 1972. DM 24,-

Vol. 294: Stability of Stochastic Dynamical Systems. Edited by R. F. Curtain. IX, 332 pages. 1972. DM 26,-

Vol. 295: C. Dellacherie, Ensembles Analytiques, Capacités, Mesures de Hausdorff. XII, 123 pages. 1972. DM 16,-

Vol. 296: Probability and Information Theory II. Edited by M. Behara, K. Krickeberg and J. Wolfowitz. V, 223 pages. 1973. DM 20,-

Vol. 297: J. Garnett, Analytic Capacity and Measure. IV, 138 pages. 1972. DM 16,-

Vol. 298: Proceedings of the Second Conference on Compact Transformation Groups. Part 1. XIII, 453 pages. 1972. DM 32,-

Vol. 299: Proceedings of the Second Conference on Compact Transformation Groups. Part 2. XIV, 327 pages. 1972. DM 26,-

Vol. 300: P. Eymard, Moyennes Invariantes et Représentations Unitaires. II. 113 pages. 1972. DM 16,-

Vol. 301: F. Pittnauer, Vorlesungen über asymptotische Reihen. VI, 186 Seiten. 1972. DM 18,-

Vol. 302: M. Demazure, Lectures on p-Divisible Groups. V, 98 pages. 1972. DM 16,-

Vol. 303: Graph Theory and Applications. Edited by Y. Alavi, D. R. Lick and A. T. White. IX, 329 pages. 1972. DM 26,-

Vol. 304: A. K. Bousfield and D. M. Kan, Homotopy Limits, Completions and Localizations. V, 348 pages. 1972. DM 26,-

Vol. 305: Théorie des Topos et Cohomologie Etale des Schémas. Tome 3. (SGA 4). Dirigé par M. Artin, A. Grothendieck et J. L. Verdier. VI, 640 pages. 1973. DM 50,-

Vol. 306: H. Luckhardt, Extensional Gödel Functional Interpretation. VI, 161 pages. 1973. DM 18,-

Vol. 307: J. L. Bretagnolle, S. D. Chatterji et P.-A. Meyer, Ecole d'été de Probabilités: Processus Stochastiques. VI, 198 pages. 1973. DM 20,-

Vol. 308: D. Knutson, λ-Rings and the Representation Theory of the Symmetric Group. IV, 203 pages. 1973. DM 20,-

Vol. 309: D. H. Sattinger, Topics in Stability and Bifurcation Theory. VI, 190 pages. 1973. DM 18,-

Vol. 310: B. Iversen, Generic Local Structure of the Morphisms in Commutative Algebra. IV, 108 pages. 1973. DM 16,-

Vol. 311: Conference on Commutative Algebra. Edited by J. W. Brewer and E. A. Rutter. VII, 251 pages. 1973. DM 22,-

Vol. 312: Symposium on Ordinary Differential Equations. Edited by W. A. Harris, Jr. and Y. Sibuya. VIII, 204 pages. 1973. DM 22,-

Vol. 313: K. Jörgens and J. Weidmann, Spectral Properties of Hamiltonian Operators. III, 140 pages. 1973. DM 16,-

Vol. 314: M. Deuring, Lectures on the Theory of Algebraic Functions of One Variable. VI, 151 pages. 1973. DM 16,-

Vol. 315: K. Bichteler, Integration Theory (with Special Attention to Vector Measures). VI, 357 pages. 1973. DM 26,-

Vol. 316: Symposium on Non-Well-Posed Problems and Logarithmic Convexity. Edited by R. J. Knops. V, 176 pages. 1973. DM 18,-

Vol. 317: Séminaire Bourbaki – vol. 1971/72. Exposés 400–417. IV, 361 pages. 1973. DM 26,-

Vol. 318: Recent Advances in Topological Dynamics. Edited by A. Beck, VIII, 285 pages. 1973. DM 24,-

Vol. 319: Conference on Group Theory. Edited by R. W. Gatterdam and K. W. Weston. V, 188 pages. 1973. DM 18,-

Vol. 320: Modular Functions of One Variable I. Edited by W. Kuyk. V, 195 pages. 1973. DM 18,-

Vol. 321: Séminaire de Probabilités VII. Edité par P. A. Meyer. VI, 322 pages. 1973. DM 26,-

Vol. 322: Nonlinear Problems in the Physical Sciences and Biology. Edited by I. Stakgold, D. D. Joseph and D. H. Sattinger. VIII, 357 pages. 1973. DM 26,-

Vol. 323: J. L. Lions, Perturbations Singulières dans les Problèmes aux Limites et en Contrôle Optimal. XII, 645 pages. 1973. DM 42,-

Vol. 324: K. Kreith, Oscillation Theory. VI, 109 pages. 1973. DM 16,-

Vol. 325: Ch.-Ch. Chou, La Transformation de Fourier Complexe et L'Equation de Convolution. IX, 137 pages. 1973. DM 16,-

Vol. 326: A. Robert, Elliptic Curves. VIII, 264 pages. 1973. DM 22,-

Vol. 327: E. Matlis, 1-Dimensional Cohen-Macaulay Rings. XII, 157 pages. 1973. DM 18,-

Vol. 328: J. R. Büchi and D. Siefkes, The Monadic Second Order Theory of All Countable Ordinals. VI, 217 pages. 1973. DM 20,-

Vol. 329: W. Trebels, Multipliers for (C, α)-Bounded Fourier Expansions in Banach Spaces and Approximation Theory. VII, 103 pages. 1973. DM 16,-

Vol. 330: Proceedings of the Second Japan-USSR Symposium on Probability Theory. Edited by G. Maruyama and Yu. V. Prokhorov. VI, 550 pages. 1973. DM 36,-

Vol. 331: Summer School on Topological Vector Spaces. Edited by L. Waelbroeck. VI, 226 pages. 1973. DM 20,-

Vol. 332: Séminaire Pierre Lelong (Analyse) Année 1971-1972. V, 131 pages. 1973. DM 16,-

Vol. 333: Numerische, insbesondere approximationstheoretische Behandlung von Funktionalgleichungen. Herausgegeben von R. Ansorge und W. Törnig. VI, 296 Seiten. 1973. DM 24,-

Vol. 334: F. Schweiger, The Metrical Theory of Jacobi-Perron Algorithm. V, 111 pages. 1973. DM 16,-

Vol. 335: H. Huck, R. Roitzsch, U. Simon, W. Vortisch, R. Walden, B. Wegner und W. Wendland, Beweismethoden der Differentialgeometrie im Großen. IX, 159 Seiten. 1973. DM 18,-

Vol. 336: L'Analyse Harmonique dans le Domaine Complexe. Edité par E. J. Akutowicz. VIII, 169 pages. 1973. DM 18,-

Vol. 337: Cambridge Summer School in Mathematical Logic. Edited by A. R. D. Mathias and H. Rogers. IX, 660 pages. 1973. DM 42,-

Vol. 338: J. Lindenstrauss and L. Tzafriri, Classical Banach Spaces. IX, 243 pages. 1973. DM 22,-

Vol. 339: G. Kempf, F. Knudsen, D. Mumford and B. Saint-Donat, Toroidal Embeddings I. VIII, 209 pages. 1973. DM 20,-

Vol. 340: Groupes de Monodromie en Géométrie Algébrique. (SGA 7 II). Par P. Deligne et N. Katz. X, 438 pages. 1973. DM 40,-

Vol. 341: Algebraic K-Theory I, Higher K-Theories. Edited by H. Bass. XV, 335 pages. 1973. DM 26,-

Vol. 342: Algebraic K-Theory II, "Classical" Algebraic K-Theory, and Connections with Arithmetic. Edited by H. Bass. XV, 527 pages. 1973. DM 36,-

Vol. 343: Algebraic K-Theory III, Hermitian K-Theory and Geometric Applications. Edited by H. Bass. XV, 572 pages. 1973. DM 38,-

Vol. 344: A. S. Troelstra (Editor), Metamathematical Investigation of Intuitionistic Arithmetic and Analysis. XVII, 485 pages. 1973. DM 34,-

Vol. 345: Proceedings of a Conference on Operator Theory. Edited by P. A. Fillmore. VI, 228 pages. 1973. DM 20,-

Vol. 346: Fučík et al., Spectral Analysis of Nonlinear Operators. II, 287 pages. 1973. DM 26,-

Vol. 347: J. M. Boardman and R. M. Vogt, Homotopy Invariant Algebraic Structures on Topological Spaces. X, 257 pages. 1973. DM 22,-

Vol. 348: A. M. Mathai and R. K. Saxena, Generalized Hypergeometric Functions with Applications in Statistics and Physical Sciences. VII, 314 pages. 1973. DM 26,-

Vol. 349: Modular Functions of One Variable II. Edited by W. Kuyk and P. Deligne. V, 598 pages. 1973. DM 38,-

Vol. 350: Modular Functions of One Variable III. Edited by W. Kuyk and J.-P. Serre. V, 350 pages. 1973. DM 26,-

Vol. 351: H. Tachikawa, Quasi-Frobenius Rings and Generalizations. XI, 172 pages. 1973. DM 18,-

Vol. 352: J. D. Fay, Theta Functions on Riemann Surfaces. V, 137 pages. 1973. DM 16,-

Vol. 353: Proceedings of the Conference on Orders, Group Rings and Related Topics. Organized by J. S. Hsia, M. L. Madan and T. G. Ralley. X, 224 pages. 1973. DM 20,-

Vol. 354: K. J. Devlin, Aspects of Constructibility. XII, 240 pages. 1973. DM 22,-

Vol. 355: M. Sion, A Theory of Semigroup Valued Measures. V, 140 pages. 1973. DM 16,-

Vol. 356: W. L. J. van der Kallen, Infinitesimally Central-Extensions of Chevalley Groups. VII, 147 pages. 1973. DM 16,-

Vol. 357: W. Borho, P. Gabriel und R. Rentschler, Primideale in Einhüllenden auflösbarer Lie-Algebren. V, 182 Seiten. 1973. DM 18,-

Vol. 358: F. L. Williams, Tensor Products of Principal Series Representations. VI, 132 pages. 1973. DM 16,-

Vol. 359: U. Stammbach, Homology in Group Theory. VIII, 183 pages. 1973. DM 18,-

Vol. 360: W. J. Padgett and R. L. Taylor, Laws of Large Numbers for Normed Linear Spaces and Certain Fréchet Spaces. VI, 111 pages. 1973. DM 16,-

Vol. 361: J. W. Schutz, Foundations of Special Relativity: Kinematic Axioms for Minkowski Space Time. XX, 314 pages. 1973. DM 26,-

Vol. 362: Proceedings of the Conference on Numerical Solution of Ordinary Differential Equations. Edited by D. Bettis. VIII, 490 pages. 1974. DM 34,-

Vol. 363: Conference on the Numerical Solution of Differential Equations. Edited by G. A. Watson. IX, 221 pages. 1974. DM 20,-

Vol. 364: Proceedings on Infinite Dimensional Holomorphy. Edited by T. L. Hayden and T. J. Suffridge. VII, 212 pages. 1974. DM 20,-

Vol. 365: R. P. Gilbert, Constructive Methods for Elliptic Equations. VII, 397 pages. 1974. DM 26,-

Vol. 366: R. Steinberg, Conjugacy Classes in Algebraic Groups (Notes by V. V. Deodhar). VI, 159 pages. 1974. DM 18,-

Vol. 367: K. Langmann und W. Lütkebohmert, Cousinverteilungen und Fortsetzungssätze. VI, 151 Seiten. 1974. DM 16,-

Vol. 368: R. J. Milgram, Unstable Homotopy from the Stable Point of View. V, 109 pages. 1974. DM 16,-

Vol. 369: Victoria Symposium on Nonstandard Analysis. Edited by A. Hurd and P. Loeb. XVIII, 339 pages. 1974. DM 26,-

Vol. 370: B. Mazur and W. Messing, Universal Extensions and One Dimensional Crystalline Cohomology. VII, 134 pages. 1974. DM 16,-

Vol. 371: V. Poenaru, Analyse Différentielle. V, 228 pages. 1974. DM 20,-

Vol. 372: Proceedings of the Second International Conference on the Theory of Groups 1973. Edited by M. F. Newman. VII, 740 pages. 1974. DM 48,-

Vol. 373: A. E. R. Woodcock and T. Poston, A Geometrical Study of the Elementary Catastrophes. V, 257 pages. 1974. DM 22,-

Vol. 374 S. Yamamuro, Differential Calculus in Topological Linear Spaces. IV, 179 pages. 1974. DM 18,-

Vol. 375: Topology Conference 1973. Edited by R. F. Dickman Jr. and P. Fletcher. X, 283 pages. 1974. DM 24,-

Vol. 376: D. B. Osteyee and I. J. Good, Information, Weight of Evidence, the Singularity between Probability Measures and Signal Detection. XI, 156 pages. 1974. DM 16.-

Vol. 377: A. M. Fink, Almost Periodic Differential Equations. VIII, 336 pages. 1974. DM 26,-

Vol. 378: TOPO 72 – General Topology and its Applications. Proceedings 1972. Edited by R. Alò, R. W. Heath and J. Nagata. XIV, 651 pages. 1974. DM 50,-

Vol. 379: A. Badrikian et S. Chevet, Mesures Cylindriques, Espaces de Wiener et Fonctions Aléatoires Gaussiennes. X, 383 pages. 1974. DM 32,-

Vol. 380: M. Petrich, Rings and Semigroups. VIII, 182 pages. 1974. DM 18,-

Vol. 381: Séminaire de Probabilités VIII. Edité par P. A. Meyer. IX, 354 pages. 1974. DM 32,-

Vol. 382: J. H. van Lint, Combinatorial Theory Seminar Eindhoven University of Technology. VI, 131 pages. 1974. DM 18,-

Vol. 383: Séminaire Bourbaki – vol. 1972/73. Exposés 418-435. IV, 334 pages. 1974. DM 30,-

Vol. 384: Functional Analysis and Applications, Proceedings 1972. Edited by L. Nachbin. V, 270 pages. 1974. DM 22,-

Vol. 385: J. Douglas Jr. and T. Dupont, Collocation Methods for Parabolic Equations in a Single Space Variable (Based on C^1-Piecewise-Polynomial Spaces). V, 147 pages. 1974. DM 16,-

Vol. 386: J. Tits, Buildings of Spherical Type and Finite BN-Pairs. IX, 299 pages. 1974. DM 24,-

Vol. 387: C. P. Bruter, Eléments de la Théorie des Matroïdes. V, 138 pages. 1974. DM 18,-

Vol. 388: R. L. Lipsman, Group Representations. X, 166 pages. 1974. DM 20,-

Vol. 389: M.-A. Knus et M. Ojanguren, Théorie de la Descente et Algèbres d' Azumaya. IV, 163 pages. 1974. DM 20,-

Vol. 390: P. A. Meyer, P. Priouret et F. Spitzer, Ecole d'Eté de Probabilités de Saint-Flour III – 1973. Edité par A. Badrikian et P.-L. Hennequin. VIII, 189 pages. 1974. DM 20,-

Vol. 391: J. Gray, Formal Category Theory: Adjointness for 2-Categories. XII, 282 pages. 1974. DM 24,-

Vol. 392: Géométrie Différentielle, Colloque, Santiago de Compostela, Espagne 1972. Edité par E. Vidal. VI, 225 pages. 1974. DM 20,-

Vol. 393: G. Wassermann, Stability of Unfoldings. IX, 164 pages. 1974. DM 20,-

Vol. 394: W. M. Patterson 3rd, Iterative Methods for the Solution of a Linear Operator Equation in Hilbert Space – A Survey. III, 183 pages. 1974. DM 20,-

Vol. 395: Numerische Behandlung nichtlinearer Integrodifferential- und Differentialgleichungen. Tagung 1973. Herausgegeben von R. Ansorge und W. Törnig. VII, 313 Seiten. 1974. DM 28,-

Vol. 396: K. H. Hofmann, M. Mislove and A. Stralka, The Pontryagin Duality of Compact O-Dimensional Semilattices and its Applications. XVI, 122 pages. 1974. DM 18,-

Vol. 397: T. Yamada, The Schur Subgroup of the Brauer Group. V, 159 pages. 1974. DM 18,-

Vol. 398: Théories de l'Information, Actes des Rencontres de Marseille-Luminy, 1973. Edité par J. Kampé de Fériet et C. Picard. XII, 201 pages. 1974. DM 23,-